CONSTRUCTION ESTIMATING

PROFESSIONAL REFERENCE

Adam Ding

Created exclusively
for DeWALT by:

www.palpublications.com
1-800-246-2175

Titles Available From DeWALT

DeWALT Trade Reference Series

Blueprint Reading Professional Reference
Construction Professional Reference
Construction Estimating Professional Reference
Electric Motor Professional Reference
Electrical Estimating Professional Reference
Electrical Professional Reference
HVAC/R Professional Reference – Master Edition
Lighting & Maintenance Professional Reference
Plumbing Professional Reference
Spanish/English Construction Dictionary – Illustrated
Wiring Diagrams Professional Reference

DeWALT Exam and Certification Series

Construction Licensing Exam Guide
Electrical Licensing Exam Guide
HVAC Technician Certification Exam Guide
Plumbing Licensing Exam Guide

For a complete list of The DeWALT Professional Trade Reference Series visit **www.palpublications.com**.

This Book Belongs To:

Name:_____

Company: _____

Title: _____

Department: _____

Company Address: _____

_____

_____

Company Phone: _____

Home Phone: _____

Pal Publications, Inc.
800 Heritage Drive, Suite 810
Pottstown, PA 19464-3810

NOTICE OF RIGHTS

NOTICE OF LIABILITY

ISBN 0-9777183-0-1

10 09 08 07 06   5 4 3 2 1

Printed in the United States of America

A Note To Our Customers

We have manufactured this book to the highest quality standards possible. The cover is made of a flexible, durable and water-resistant material able to withstand the toughest on-the-job conditions. We also utilize the Otabind process which allows this book to lay flatter than traditional paperback books that tend to snap shut while in use.

Preface

Bidding to get a job can be tough sometimes. An estimator who has the success rate over 10% might deserve a raise. You want to be the low bidder. But you do not want to be so low that making a reasonable profit is almost impossible.

This book is written to help you to better organize a busy life, win more bids, and make fewer mistakes. today there are lots of books available to show you how to do quantity take-offs, but the ones focusing on winning a bid seem rare. This is why you are reading this book.

Based on the real-life experience in bidding hundreds of different jobs, this book walks you through every step of the bidding process — from the moment you receive a set of drawings to the post-bid review. The emphasis is on winning a job, sooner or later, while minimizing the errors. A number of checklists, forms, and "how-to" instructions are included. You will be able to use them on your current bid immediately.

Please read on and you will open a new window to bidding success!

Good Luck.
Adam Ding

CONTENTS

Chapter 2 – *Estimating Trades* 2-1

CHAPTER 3 – *Bid Day* 3-1

Chapter 7 – *Trade Estimating Forms* . . 7-1

CHAPTER 1
General Preparation

Successful construction bidding is a "lost art." As an estimator, your success in biddings often determines how much business your company will receive. But, in reality, a success rate of more than 10% is rare. Someone once said, "We do not just expect miracles; we depend on them."

OVERALL ORGANIZATION

To improve their chances of success, people often bid 5 to 6 jobs at the same time, hoping to get at least one. Each bid costs time and money. To make your efforts effective, you need to have priorities and good organization.

Having priorities means focusing more on bids for jobs that you are likely to win and make a profit from. Use the worksheet in **Figure 1.1** to help you decide which bids deserve your immediate focus. The following are factors to consider:

1. Date: Which bid is due first, second, and so on? Normally you would first work on bids that are due earlier. If the dates for the jobs you are bidding on are so close that you will not have time to do more bids during those dates, then you might consider spending a little more time now on bids that are due later.

2. Size: How big is each job? How many square feet is the building? How many units if residential development? How many rooms in the hotel? How many students for the school? How many beds in the hospital? How many acres is the site? How many parking spaces? What's the approximate total dollar value for the job? Usually the bigger the job, the more effort required to bid and build, but also there is more profit to be made.

3. Type of construction: How complex is each bid? What types of systems for foundation, wall, and roof? Is there anything special about the job, like auger piling? Although a complex job takes more effort and means more risks, it could be more profitable.

4. Location: How far is each job from your office? Having a job close to your work area means cost savings in overhead.

5. Line of expertise: Which job you are more familiar with? You may have completed a few jobs like this before, and that experience could help you get this one.

6. Performing capacity: How many jobs do you currently have under way? Among them, how many are near completion? If another job is awarded, will it exceed your bonding capacity? Will your staff be available to run the job to finish it on time? You do not want to win a job you will not have the capacity to complete efficiently.

7. Competition: How many other contractors are you bidding against? Did you bid against them before? Did you win? Normally it is easier to win a bid with less competition.

8. Owner: Does the owner have enough money to finance the job? How is their reputation? How and when you will expect to receive the payments?

9. Designer: Who is the architect? Who is the engineer? How is their reputation? Do they produce good quality drawings? Sketchy documents could later result in numerous change orders and associated disputes.

For each bid, it is important to have a clean approach to organize the work. Take a good look at **Figure 1.2**. To "beat the clock," you will need to get several things going at the same time. By following the work flow chart given, you could save significant time and effort.

REVIEWING BIDDING DOCUMENTS

To set everything straight from the very beginning, you need to have a good review of the bidding documents.

1. Read *Instruction To Bidders*

Bid instructions could be included as part of the spec book, or as a separate document. These instructions talk about the procedures to submit proposals. Read every word in this important document, and use **Figure 1.3** to summarize your findings.

By completing a pre-bid checklist, you will always have the "big-picture" information handy.

2. Check package completeness

Use the worksheet in **Figure 1.4** to make a quick check to ensure the package is complete. Not all of the listed documents are expected to exist on every project. For example, in a small warehouse that is not air-conditioned, you won't need HVAC drawings. Also, please note that some documents could be mixed with others. For example, specifications, usually appearing as a separate book, could be laid out on the drawing package like normal sheets.

3. Study drawings

First take a count of drawings to make sure you have them all. Find the drawing index, which is a comprehensive list of all the drawings included in the bid. Then flip through each individual drawing sheet and check sheet numbers off the drawing index. Make a note of missing sheets or extra sheets.

Now thoroughly go through one sheet after another to get the information you need. All the sheets, cross-referencing each other, should be reviewed. Use the question list in **Figure 1.5** to help you. While reviewing drawings, circle unfamiliar details with a red pencil so you can refer to them easily later.

4. Examine specs

First find the table of contents in the spec book, and take note of any missing sections. Start by reading the sections describing general aspects of the job. Pay attention to supplementary or special conditions which cover the specific requirements for this particular job. You should make a list of items affecting costs such as working hours, project access, construction parking, noise abatement, material substitutions, change orders, conditions of final payment, etc.

Next go over all sections in the spec book for different trades. Find out what kind of materials or systems are to be used. Try to match the building components shown on drawings with their requirements in specs. Because most technical specifications today use the CSI-division (Construction Specification Institute) format, you can use the list of trade sections in the spec book to help define the general bid scope.

5. Make Notes

A good estimator always makes detailed notes. The specific format of your notes could make a difference. Use the worksheet in **Figure 1.6** to record each note according to which trade it applies and where it can be found. At any point of the reviewing process, you can go back to this worksheet, adding notes for any trade. Finally, you should have a long list of items affecting the costs. **Figure 1.7** shows a completed example.

REVIEWING BIDDING DOCUMENTS *(cont.)*

Reviewing documents is more than just "taking-off" the information to your note pad. You should think more about what it takes to construct the building. Not all information is shown on drawings. You are the one to build the job, not the architect, so try to visualize how you would put all of the components together to form a building.

What you get from reviewing documents is the basis for further actions. For instance, you may need to arrange for someone to deliver the bid if it requires personal delivery, mark your calendar to attend pre-bid meeting, request a quote for bond and insurance, contact the architect to get clarifications on some unclear issues, etc. See **Figure 1.8** for an example of a bid RFI to an architect.

CONTACTING SUBS AND SUPPLIERS

Using subcontractors on a job is common. An electrical contractor may seem like a sub to a general contractor, while a data communication contractor could be the sub for that electrical contractor. Sometimes people hire subs even if they are licensed to do the work themselves, because subs can be more cost-effective.

If you need subs or suppliers to bid a job, it is best to notify them as early as possible. If delayed, subs either can not finish the bid or just bid high to keep themselves covered. This could make your total bid high, and you could lose the deal.

1. Write a bid invitation.

A bid invitation, the official document containing general bid information, should be created to communicate with your subs and suppliers. A well-written invitation saves hours of explanation on the phone. Use **Figure 1.9** as a template. You should have reviewed drawings and specs before writing an invitation. After your review, you should have a clear idea of what the job entails.

Bid invitations could be sent to subs or suppliers in a number of ways, such as fax, mail, or email; but they should only be sent to the subs whose work is related to the bid.

2. Make documents available.

Nobody can bid a job well without having reviewed the documents. You could store the documents with a reprographics company and inform your subs or suppliers that plans are available from there. Try to have clear communication with the reprographics people, especially if you want them to keep track of further changes to the documents.

Organize drawings before sending them off to make copies. For example, you can combine on-site and off-site drawings and mark them as "Civil Drawings"; combine landscaping and irrigation drawings if they are separate; mark packages as "Building A plans", "Building B plans," etc.

Some subs or suppliers don't want to pay for drawings to bid jobs. You may consider setting aside a work space in your company where subs could come and view plans. Make sure subs have access to the whole set of documents, not just portions of plans. Otherwise, they might turn in incorrect or incomplete bids and claim they lacked information. For example, if you only provide a concrete sub with the architectural and structural sheets, he might exclude the concrete work associated with plumbing trenches and electrical conduits.

By having subs review plans in your office, you can interact with them, talk through problems, and encourage more competitive prices. For example, if a sub thinks the flooring material specified has been discontinued; you could talk with him right away and communicate the issue back to the architect.

3. Use qualified sub.

Finding the right subs is not just for the sake of bidding. If during construction, one sub can't perform, then you have to find a replacement and pay for the price difference and associated delays.

A popular way to identify qualified subs is by word-of-mouth. Owners, architects, and engineers usually have a few subs they like to work with. Your local trade associations or construction reporting services could also recommend a list of their premium members.

Another good way to find qualified and dependable subs is to contact material suppliers for a list of people they recommend. For example:

    a. Contact pre-cast storm structure suppliers to find site subs.

    b. Contact ready mix suppliers to find concrete subs.

CONTACTING SUBS AND SUPPLIERS *(cont.)*

   c. Contact block, brick, or rebar suppliers to find masonry subs.

   d. Contact joist and decking suppliers to find steel subs.

   e. Contact HVAC unit suppliers to find mechanical subs.

   f. Contact lighting fixture or switchgear suppliers to find electrical subs.

The names of some subs may not be familiar to you. To have peace of mind, you can "screen" them by asking for a list of professional references from the jobs they have done before. Check out each reference. You could also ask them to provide supporting documents such as insurance certificates, business licenses, etc.

SITE INVESTIGATION

Before sending a bid, you should always visit the site. Site conditions might be different from what you see on drawings. Even if the architect did a good job in showing what is there, it is recommended to visit the actual site to get a better idea.

A site investigation could be done step by step as follows:

1. Before you go

How much do you know? Get familiar with drawings and specs. Pay special attention to documents describing existing conditions, such as the soil report, demolition plan, boundary survey, etc. Write down your questions and try to look for answers when you get there.

How to get there? Find out the exact job location and the quickest way to get there. You might need a detailed map or driving directions. Some good references could be found on the Internet. Keep whatever you find.

Who else is going? You may want to invite your field personnel to come along.

Ask your project managers or construction superintendents if they are free. You might also consider inviting some good subs. Of course, contact them several days in advance to make sure they are available. Bringing these people may help you identify critical cost issues, and this could improve your chances of quoting the job accurately.

What to bring along? Bring a digital camera, and remember to take a set of drawings and specifications along to the site. You might also consider bringing a video camcorder, measuring tape, and other necessary aids like an earth auger and a hand level. Chances are you will need them.

2. On the road

On the way to the site, look for road conditions, weight and height limits of bridges leading to the job, etc. If the site is hard to access, problems may arise in getting material and equipment to the site after the job begins.

3. On the site

a. Which way is North? As soon as you are on site, make a sketch of site layout and proposed building orientation.

b. What's already there? Walk through the site. Inspect existing conditions and compare them with what is on the drawings. Visualize how the building could be put together on site. Look for answers to questions you already had in mind.

c. Get down to earth. Examine existing soil conditions, and compare them with the soil report. Is the soil strong enough to support the proposed structure? Is there any hazardous material present? Could groundwater require pumping during construction? Does the site drain well if it rains? Use the earth auger you brought to find out.

d. Locate utilities. Find connections for existing electricity, water, telephone, sanitary sewer, etc. Determine temporary utilities required for construction. For example, you might have to rent a generator to provide enough power. Also decide future locations for trash dumpsters, material storage trailers, portable lavatories, etc.

e. What about existing structures? If there's an existing building, try to gain access and walk through it. Is the building to be demolished, expanded, or renovated? If demolished, are there any hazardous materials present that will hold up the demolition? If expanded or renovated, are there any requirements on matching existing materials and finishes? Is the existing building to be kept in operation while the new construction is going on? If so, are there any requirements to control dust and noise, such as limited working hours? Can you get heavy equipment into the building?

f. Who's next door? Look around the site. Is it in a safe neighborhood? You might need some security systems in place such as fences and gates to protect the construction material as well as the crew. Walk around adjacent structures to see if they will need temporary protection or underpinning once construction starts.

g. Cooperate with authorities. Visit the local building department that has jurisdiction in the area. Learn about building ordinances, regulations, labor conditions, crimes, availability of housing, etc.

h. Take notes. Remember to take as many notes as you can of your findings. Technologies like digital cameras have made taking pictures very inexpensive, so remember to take a lot of pictures, too.

4. Back to office

After you've made it back to the office, download the pictures, and summarize your findings in an investigation report. Review the information, and ask the owner or architect for clarification on some issues if necessary. If you feel that what you have is not enough, arrange follow-up visits to acquire more information.

Site visits are very important, especially when you are bidding an unfamiliar, complex job in a remote location. Sometimes as the deadline approaches, and the site is far from the office, you may have no time to do a site visit. If so, another experienced person could visit and prepare a site investigation report for you. Use Figure 1.10 as a guideline.

FIGURE 1.1 — PLANNING MULTIPLE BIDS

Evaluation Factors	Bid #1	Bid #2	Bid #3	Bid #4	Bid #5
Job Name					
Bid Date					
Size					
Type of Construction					
Location					
Similar to previous jobs?					
Within performing capacity?					
Number of competitors					
Owner's reputation					
Designer's reputation					
Overall Winning Chances					

FIGURE 1.2 — BID WORK FLOW

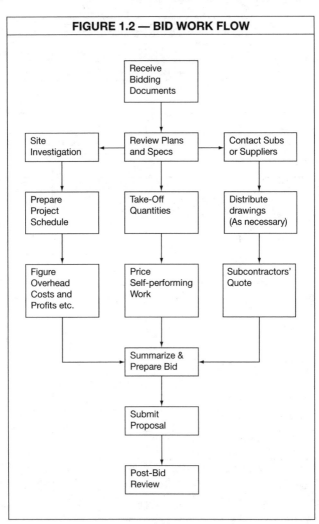

FIGURE 1.3 — PRE-BID CHECKLIST

Job Name: _____

Bid Due Date/Time:_____

Pre-bid Meeting:
Date: _____   Time:_____
Location: _____   Participants: _____

Project   Project
Start Date: Finish Date: _____

Bid Delivery
Format:
❑ Fax ❑ E-mail ❑ Personal ❑ Courier

Send Bid To:
❑ Architect
❑ Owner, Contact Info: _____

_____

_____

Job Location:_____

Tax Rate: _____

Construction Type: _____

Building Area: _____   Site Acres:_____

Approximate Value: _____

FIGURE 1.3 — PRE-BID CHECKLIST *(cont.)*

Owner's Name: _____

Architect's Name: _____

Names of Competition: _____

Liquidated Damages: _____

Labor Conditions: ❑ Prevailing wages ❑ Union
                    ❑ Open Shop

Permit Costs: ❑ Include in bid proposal
             ❑ Exclude (By Owner)

Development Fees: ❑ Include in bid proposal
                    ❑ Exclude (By Owner)

Bond: ❑ Bid bonds ❑ Payment and Performance bonds

Insurance: ❑ Builder's Risk ❑ Liability Insurance
            ❑ Owner's Protective Insurance

Material Testing ❑ Include in bid proposal
                ❑ Exclude (By Owner)

Work Done by Owner's Contractor
(Exclude from the bid): _____

Cost breakdown requirements: _____

Alternates: _____

FIGURE 1.4 — CHECKING PACKAGE COMPLETENESS

Job Name: _____

Checked By: _____

❏ Civil Drawings
  ❏ Survey
  ❏ Demolition
  ❏ Paving
  ❏ Utilities
  ❏ Site Electrical
  ❏ Soil Report
  ❏ Landscaping and Irrigation

❏ Architectural Drawings
  ❏ Floor Plan
  ❏ Exterior/Interior Elevation
  ❏ Roof Plan and Reflected Ceiling Plan
  ❏ Sections and Details
  ❏ Interior Design

❏ Structural drawings
  ❏ Foundation
  ❏ Floor Framing
  ❏ Roof Framing
  ❏ Sections and Details

❏ Mechanical and Electrical
  ❏ Fire Protection
  ❏ HVAC
  ❏ Plumbing
  ❏ Electrical

❏ Specifications Book

FIGURE 1.5 — QUESTION LIST FOR REVIEWING DOCUMENTS

Site Drawings and Specs

- Is it an existing site or new site?

- Is there a lot of earthwork to be done?

- Is there any offsite roadwork?

- Is the building to be serviced with complete utilities like water and sewer?

- What are soil conditions according to the geo-tech report?

- Should the excavated material be hauled from the site? Can it be re-used?

- Can you burn trees on site?

Architectural Drawings and Specs

- Is there an existing building to be demolished, renovated, or expanded?

- What shape is the new building?

- How many square feet are the new building and renovated building combined?

- How long is the perimeter of the new building?

- How many floors, and what's on each floor?

- Where is the building entrance?

- What are the names of each of the rooms, and how big are they?

- How high is exterior wall and what is it made of?

FIGURE 1.5 — QUESTION LIST FOR REVIEWING DOCUMENTS *(cont.)*

- What are the principal exterior finish materials?

- What types of doors and windows do you have, and is there a schedule?

- What types of roof systems are specified?

- Are there any balconies, canopies, or walkways?

- How high is the interior ceiling?

- What's the partition wall made of, and is it load-bearing?

- What's the floor finish material, and is there a finish schedule?

- Is the building interior to be painted?

- Are there any elevators, stairs, rails, fireplaces, or chimneys?

- What types of fixtures or equipment are you required to furnish and install?

Structural Drawings and Specs

- What types of systems are used?

- How deep is the foundation, and is there any special drainage system?

- Is the foundation wall cast-in-place concrete or masonry?

- What's the concrete strength required?

- How high is the roof?

**FIGURE 1.5 — QUESTION LIST FOR
REVIEWING DOCUMENTS** *(cont.)*

- Is there any floor below grade, and how deep is it?

- Are the floors made of concrete, steel, or wood?

- What is the roof structure made of?

- How is the structure fire-protected?

- Will any heavy construction equipment such as a crane be required to do the work?

Mechanical and Electrical Drawings and Specs

- What types of systems are used?

- What types of materials are specified for equipment, piping, wiring, or fixtures?

- What work will be done by public utility companies?

- How will mechanical and electrical systems affect the site construction?

- How will mechanical and electrical systems affect the building construction?

FIGURE 1.6 — WORKSHEET FOR MAKING REVIEW NOTES

Job Name:

Notes Made By:

Date:

Trades	Drawing Sheets or Spec Sections	Review Notes
Site		
Concrete		
Masonry		
Metals		
Wood/Plastics		
Thermal/Moisture Protection		
Doors and Windows		
Finishes		
Specialties		
Equipment		
Furnishings		
Mechanical		
Electrical		

FIGURE 1.7 — COMPLETED REVIEW NOTES

Job Name: All American School	Notes Made By: AD	Date: January 1, 2005
Trades	Reference	Review Notes
Sitework	C-1	Certified Survey and As-Builts
		Material Testing to be Included
	C-2	Demo Existing Building
	C-3	Temp Access Road Required
	C-4	Offsite work and Retention Pond
		MOT and Traffic Signalization
	RFI	Where's soil report? Existing Survey?
	C-5	Site Fire Line and DDCV
		Need to Send Plans to ABC Sitework
Landscape/Irrigation	RFI	Owner's Allowance? Need Clarification
	C-2	Existing Tree Relocation
	L-1	Need plant count to compare w/Schedule
	I-1	Jack and Bore for Irrigation Conduits

Concrete	C-2	Site Dumpster and Screen Wall Footing
	C-3	Site Sidewalk, Need QTO
	A-101	Building Sidewalk
		Need to define brick paver scope
	S-201	Concrete Beams and Columns
	P-2	concrete for plumbing trenches
	E-101	concrete for electrical conduits
Masonry	C-2	Dumpster Enclosure and Site Screen Wall
	A-104	Exterior brick veneer, need contact supplier
		Need a detailed masonry QTO
Metals	C-7	Dumpster Gate and Bollards
	A-104	Stainless Steel Handrail, need quote
	A-201	Roof Ladder
		Pre-engineered Trusses, Metal/Wood?
	A-301	Support Framing for RTU
	S-100	Signed and sealed shop drawings

FIGURE 1.7 — COMPLETED REVIEW NOTES (cont.)

Job Name: All American School	Notes Made By: AD	Date: January 1, 2005
Trades	Reference	Review Notes
Wood and Plastics	A-100	Fire rated plywood for exterior canopy
	A-104	Exterior wood decking
	A-106	Millwork furnished by owner, incl. Installation
	A-107	Fypon trim, vinyl soffit
Division 7	A-105	Built-up roofing w/Hatch, tile roofing
	Specs	List of approved roofers
		Stainless steel sheet metal
		20-year warranty labor and material
	A-201	Aluminum panel on parapet wall
	RFI	Need clarification on Fireproofing
Doors	A-202	Hollow metal door, wood door

Category	Code	Description
		Overhead door, automatic door
Glass	A-202	Glass for automatic doors
	Specs	List of approved curtain wall system
	A-601	Interior Glass and Mirrors on wall
	A-104	Exterior windows
	S-100	Signed and sealed shop drawings
	RFI	Is glass supposed to be impact resistant?
Drywall/Stucco	A-201	Trusses
	S-100	Signed and sealed shop drawings
	A-601	Interior demising wall to reach roof deck
	A-104	Stucco w/EIFS trims
Ceiling/Flooring	A-603	Need flooring QTO by each type
Paint	RFI	Is interior wall to be painted?
	A-501	Vinyl wall covering

FIGURE 1.7 — COMPLETED REVIEW NOTES (cont.)

Job Name: All American School	Notes Made By: AD	Date: January 1, 2005
Trades	Reference	Review Notes
Division 10-14	A-303	Fire extinguishers to be included
	C-2	Site chain link fence
	A-401	Metal awnings
	A-601	Ceiling access panels
	A-101	Mailboxes, trash receptacles, bike racks
HVAC	M-1	Sheet metal ductwork for base bid
	M-2	Incl. condensate drainage piping
	RFI	Owner furnishing RTU?
Fire	C-5	Site Fire Line
	RFI	Need fire sprinkler drawings

		Is fire protection req'd for exterior canopy?
Plumbing	C-5	Two Grease Traps
	P-1	Incl. water heater
	A-105	Roof drainage connection
Electrical	C-2	Power for Monument signs
	C-5	Power for Lift Station
	I-1	Power for Irrigation
	E-100	Fire alarm system required
	RFI	Are panels and fixtures to be supplied by owner?
	Specs	Incl. Temp power
	SE-100	Incl. concrete bases for lighting poles

FIGURE 1.8 — RFI TO ARCHITECT

Request For Information #1
All American School
Date: January 1st, 2005

To: ABC Architects
Tel: (999) 123-4567
Fax: (999) 123-4568

From: Estimator
XYZ Contractor
Tel: (111) 123-4567
Fax: (111) 123-4568

1. Please provide soil report if available.

2. Please clarify whether landscape and irrigation should be included in our proposal.

3. Please confirm that the glass is to be impact resistant.

4. Please confirm that the interior wall is to be painted.

5. Please provide fire sprinkler drawings if available.

Feel free to call if you have any questions.

FIGURE 1.9 — BID INVITATION WORKSHEET

Invitation to Bid

Project name: _____

Project address: _____

_____

_____

Pricing due date and time: _____

Scope of work: _____

Plans and specs are available at: _____

Cost breakdown: _____

Alternates: _____

Bid contact: _____

Your feedback: _____

FIGURE 1.10 — SITE VISIT WORKSHEET

Date: _____

By: _____

Job name: _____

Address: _____

Directions: _____

Site access: _____

Site description: _____

Soil conditions: _____

Water table and drainage: _____

Hazardous material: _____

Availability of utilities
(water, sewer, electricity): _____

If existing building, describe situation: _____

Neighborhood information: _____

Local building codes: _____

Situation with subs: _____

Labor conditions: _____

Equipment to be used: _____

Comments: _____

Photos attached: _____

CHAPTER 2
Estimating Trades

Accurate estimating is very important to every trade you self-perform. Even if you are subcontracting the work, preparing a detailed estimate will make you more familiar with its scope. The trade estimating process is usually divided into two parts: quantity take-off and pricing.

QUANTITY TAKE-OFF

To "take-off" a job means that you "take" the information "off" the documents and translate it into a list of items with quantities.

There are three steps to a quantity take-off:

1. Define take-2-5 scope. Examine plans and specs to see what you need to take-off for this trade. For unclear details, you should ask the architect instead of making assumptions. At the end of this chapter, you will find some checklists for estimating common construction trades. Please note that you should carefully evaluate the scope of each individual job while using these checklists. Not all items in these checklists are applicable, nor they are all-inclusive.

2. Measure each item. Use dimensions specified, and do not scale drawings unless absolutely necessary. Mark drawings for the items you took off.

Because you might never be able to finish a quantity take-off without interruptions, marking drawings reminds you of what has been taken-off. This helps you avoid double counting or missing items.

3. Record quantities. Make detailed reference of which sheet you found each item on and where items exist in the building. Record your quantities with drawing number, detail number, and grid reference. It is important to keep all items separate.

a. Conversions

You need to know some basic rules in converting units and numbers. **Figure 2.1** offers information on converting between different units in length, area, volume, and weight. Always think about unit conversion in quantity take-off.

For example, concrete needed to pour a 4" slab 70' long and 50' wide can be figured as (net quantities): Volume = Length × Width × Thickness = 70' × 50' × 4"/12 = 1,167 CF or 1,167/27= 43.2 CY.

For easier calculation, convert all lengths to linear feet in decimal format. Refer to **Figure 2.2** for converting inches to feet. Sometimes you need to convert the fractions of the inches into feet such as $\frac{1}{2}$", $\frac{5}{8}$" etc. Just remember $\frac{1}{8}$" = 0.01 foot, and go from there. For example, to convert $9\frac{3}{8}$" to feet, from the table given in **Figure 2.2**, you know 9" is 0.75;. Then $\frac{3}{8}$"= 3 × 0.01 = 0.03'. Therefore $9\frac{3}{8}$" is 0.75'+ 0.03' = 0.78'.

b. Quantity Take-Off Forms

Because you take so much information from drawings, it is best to use some kind of forms to be organized. **Figure 2.3** shows a common "quantity take-off work sheet" that you could customize for specific trade needs. **Figure 2.4** shows a finished carpentry take-off using modified sheet. Please note the math used for carpentry calculation is:

Lumber (Board Feet) =
$[$Number of Lumber (ea) $\times$ Lumber Length (in feet) $\times$ Section Height (in inches) $\times$ Section Width (in inches)$] \div 12$

Plywood (square feet) =
Number of Plywood (ea) $\times$ Plywood Length (in feet) $\times$ Plywood Width (in feet)

PRICING

Pricing is the process of converting the quantity take-off into dollar values. This normally requires that you be familiar with local construction market situations. For this reason, in many construction companies, pricing is done by senior estimators.

1. Procedures

The detailed process for pricing is:

a. Summarize the quantities you have taken-off. Combine the numbers for the same types of items and allow for reasonable waste.

b. Apply material and labor unit prices to quantities.

c. Apply sales tax to material and labor burden to labor.

d. Add permits, profit, and overhead (including supervision, equipment, tools, and home office costs, etc.) to get a total price.

Pricing is typically completed using a "pricing summary sheet." Refer to **Figure 2.5** for a generic worksheet. **Figure 2.6** shows what a completed pricing summary looks like. The pricing here is based on previous quantities from **Figure 2.4** with 10% waste added.

Because pricing is more than simple math calculation, at the very least, you will need to understand where material and labor unit prices come from.

2. Material unit price

One of the best ways to get reliable material unit prices is to request quotes directly from suppliers. Make a complete list of the items you need prices for, and contact suppliers. Some suppliers may simply respond with a price list for everything they have in stock. Read quotes carefully, and verify that information such as delivery charge, sales tax, minimum orders, and expected price escalation has been stated. Sometimes suppliers will quote discounted prices to promote sales, you should be careful in deciding whether to apply the discounts.

It is important to take time to work out the right number you should use. For example, if the lumber yard is quoting by 16-foot pieces, then you need to do some conversions unless you want to take off carpentry that way. Suppose they quoted $8.00 for a 16-foot piece of two-by-four, then you can convert that to unit price by board feet (BF). A 16-foot piece of two-by-four is $[16 \times 2 \times 4] \div 12 = 10.67$ BF. So for two-by-four, material unit price is $8.00 \div 10.67$ BF = $0.75/BF.

PRICING (cont.)

3. Labor unit price

There are two factors deciding labor unit price — crew production rate and base wage rate. Look at this example: On the last job, your 3-man crew spent one 8-hour day to install toilet accessories for several bathrooms. You used 22 plywood sheets and the crew is made up of 2 carpenters and 1 helper.

2 Carpenters × 8 hours =
16 man-hours × \$50.00 =   \$800.00

1 Helper × 8 hours =
8 man-hours × \$32.00 =   \$256.00

Total: \$1, 056.00

Because they used 22 plywood sheets (4' × 8'), the sheet area is 22 × 4 × 8 = 704 SF. Divide the total wages of \$1, 056 by the sheet area, and you get \$1056 ÷ 704 = \$1.50. This is the labor unit price for installing every square foot of plywood.

In above example, the total labor spent is 16 + 8 = 24 man-hours, and each square foot of plywood takes 24 ÷ 704 = 0.034 man-hours to install. This is called crew production rate, and it is affected by a number of factors such as job conditions, location, weather, experience and morale of the team, use of construction equipment, quality of work required, etc. For example, normally a concrete block will cost less to set for its first three feet than for the balance of the wall height. That is because as the wall is built higher, each block takes longer to set.

PRICING (cont.)

The hourly rates of $50 for carpenter and $32 for helper are called base wage rates. These rates are also affected by a number of factors such as union or non-union status, normal working hours or overtime, local economy, and labor availability, etc.

To determine the labor unit price, or how much each item will cost to install, you need to review the production and payroll information from previous jobs you have completed. Apply a method similar to the one above to making your calculations.

4. Material sales tax

Please note tax is only applied to material total. The tax rate should include all taxes payable to all levels of governments (city, county, state, and federal).

5. Labor burden

Labor burden refers to the "add-on" to total hourly base wages. When a carpenter is paid $50 per hour, that $50 is the hourly base wage rate. However, the total labor costs you incur for each job is more than total base wages. The dollar difference is called labor burden.

To figure a rate for labor burden, consider a list of items including fringe benefits, social security, Federal Insurance Contributions Act (FICA), Federal Unemployment Insurance Tax (FUTA), and worker's compensation insurance, etc. The resulting rate could range from 25% to 50% of the total base wages.

ESTIMATING SYSTEMS

It is true that the best way to learn estimating is by actually doing it. But the specific method of estimating you choose could make a difference.

1. Manual estimating

This is the traditional "old-school" style estimating. You pencil down every piece of information from drawings to scratch paper, calculate all quantities by hand (perhaps with the help of a calculator), multiply them by rates, and add up everything to get a grand total price. The main drawback for such a method is that it will take a while to prepare an accurate estimate, and there are great chances of making errors and mistakes.

2. Computerized estimating

Today a lot of people are using computers in estimating to cut the time and effort. Studies show that computerized estimating can save 75% of the time of a manual estimate. Many small companies in which the owner does the estimating can use computerized estimating to allow more time for field management and business marketing.

Computerized estimates use either spreadsheet programs or specialized estimating programs.

a. Estimating using spreadsheet programs

Using spreadsheet programs in calculations makes estimates clean with fewer math errors. However, they require you to understand how formulas work and how to write them in the spreadsheet programs.

Because spreadsheet programs are not always very easy to understand, you might have a hard time making them work in the way you need. When calculations get complicated, writing correct formulas in spreadsheets becomes more difficult.

b. Estimating using specialized programs

Dozens of specialized estimating programs are available today to help with estimating and bid preparation. Many people find them easier to use than spreadsheet programs. For instance, to do a drywall estimate, you put information such as wall length, height, components, etc. into the software. With one click, the software automatically calculates all quantities for you, including stud, track, gypsum board, insulation, mud, tape, etc. Yet probably the best thing about these estimating programs is the pricing database built in them. When you've taken off quantities, the database automatically applies unit prices to all items immediately. In the database, you can store and maintain all sorts of information such as crew production rates, wage rates, equipment costs, material quotes, etc.

Since these programs are often quite expensive (normally thousands of dollars for a decent package), the question list in **Figure 2.7** could help you to make a wise purchase decision.

FIGURE 2.1 — UNIT CONVERSION

Length:	1 Foot = 12 Inches
	1 Yard = 3 Feet
Area:	1 Square Foot = 144 Square Inches
	1 Square Yard = 9 Square Feet
	1 Square = 100 Square Feet
	1 Acre = 43,560 Square Feet
Volume:	1 Cubic Yard = 27 Cubic Feet
Weight:	1 Ton = 2000 lbs

FIGURE 2.2 — CONVERTING INCHES TO FEET

Inches	In Feet
1	0.08
2	0.17
3	0.25
4	0.33
5	0.42
6	0.50
7	0.58
8	0.67
9	0.75
10	0.83
11	0.92
12	1.00

FIGURE 2.3 — QUANTITY TAKE-OFF WORKSHEET

Project:			Date:		
Location:			Page ___ of ___		
Trade:			Take-off by		

Item Description	Details				Extension
	Length	Width	Height	Count	
Total					

FIGURE 2.4 — COMPLETED CARPENTRY TAKE-OFF

Reference	Description	Lumber					Plywood					
		Qty	Grade	Length	Dimension	Ext	Qty	Grade	Thick	Length	Width	Ext
		(ea)		(ft)	(inch x inch)	(bf)	(ea)		(inch)	(ft)	(ft)	(sf)
A201	Roof Blocking/Front	2	PT	150	2X4	200						
	Roof Blocking/Sides	1	PT	100	2X6	100						
		1	PT	100	2X4	67						
	Roof Blocking/Rear	1	PT	150	2X4	100						
A202	Loading Dock Roof	2	PT	90	2X8	240						
A300	RTU Curb Blocking	2	PT	350	2X6	700						
	Roof Ladder						1	REG AC	5/8"	35	4	140
A301	Plywood/Truss Sheathing						1	FT CDX	1/2"			1625
A401	Windows/Doors Blocking	2	FT	250	1X12	500						
		1	FT	150	2X6	150						
A402	Wainscot						1	REG AC	3/4"	25	8	200
A601	Knee wall blocking	2	PT	50	2X4	67						
A602	Vanity top backing	2	FT	25	2X4	33						
	Bath Accessories Backing						1	FT CDX	1/2"			64

PT: pressure-treated, FT: fire-rated

2-13

FIGURE 2.5 — PRICING SUMMARY WORKSHEET

Item	Qty.	Unit	Material U/P	Material Subtotal	Labor U/P	Labor Subtotal	Item Subtotal
Subtotal							
Sales Tax							
Labor Burden							
Subtotal							
Overhead							
Profit							
Permit							
Total Price							

The math in pricing summary sheet is as follows:

Material Subtotal = Quantities x Material Unit Price

Labor Subtotal = Quantities x Labor Unit Price

Item Total = Material Subtotal + Labor Subtotal

Sales Tax = Material Subtotal x Tax Rate

Labor Burden = Labor Subtotal x Labor Burden Rate

Total Price = Material Subtotal + Labor Subtotal + Sales Tax + Overhead + Profit + Permit

FIGURE 2.6 — FINISHED CARPENTRY PRICING

Item	Qty.	Unit	Material U/P	Material Subtotal	Labor U/P	Labor Subtotal	Item Subtotal
PT 2X4	477	BF	$0.75	$358	$1.50	$716	$1,074
PT 2X6	880	BF	$0.85	$748	$1.50	$1,320	$2,068
PT 2X8	264	BF	$0.90	$238	$1.50	$396	$634
FT 1X12	550	BF	$1.90	$1,045	$1.50	$825	$1,870
FT 2X6	165	BF	$1.50	$248	$1.50	$248	$495
FT 2X4	36	BF	$1.45	$53	$1.50	$54	$107
REG AC 5/8"	154	SF	$1.30	$200	$1.50	$231	$431
FT CDX 1/2"	1858	SF	$1.15	$2,137	$1.50	$2,787	$4,923
REG AC 3/4"	220	SF	$1.35	$297	$1.50	$330	$627
Subtotal				**$5,323**		**$6,907**	**$12,229**
Sales Tax	7%			$373		N/A	$373
Labor Burden	30%			N/A		$2,072	$2,072
Subtotal							**$14,674**
Overhead	25%						$3,669
Profit	20%						$2,935
Total Price							**$21,277**

FIGURE 2.7 — ESTIMATING SOFTWARE EVALUATION

How does the software work for estimating?

Can the software help with the bid preparation and submission?

Does the software have features such as subcontractor and supplier management?

Does the take-off require any additional tools such as digitizers?

How does the software maintain and update material and labor prices?

What is the upfront purchase cost for the software?

Is there a charge for periodical license renewal and upgrades?

Is the software simple to use and easy to understand?

How much time can you expect to save by using the software?

Can you try the software for a period of time without purchasing?

Does the software work with any of your existing computerized systems, such as accounting and project management?

Does the software vendor provide training and technical support for his software?

How long has the software vendor been in the business?

How many customers does the vendor have?

What other software products does the vendor sell?

QUANTITY TAKE-OFF CHECKLISTS

Site Work

General

Mobilization (L/S)

Surveying and as-builts (L/S)

Site Demolition

Demolish existing building (SF or L/S)

Remove trees (EA)

Remove fence (LF)

Saw cut (LF)

Remove curb (LF)

Remove asphalt paving (SY)

Remove concrete sidewalk (SF)

Rock blasting (L/S)

Hazardous material removal (L/S)

Shoring and engineering (L/S)

Earthwork

Clear and grub (Acres)

Dewatering (L/S)

Shoring and underpinning (L/S)

Cut/Fill (CY)

Top soil removal (CY)

Place and compact (CY)

Rough grading (SY)

Soil stabilization (L/S)

Building excavation (CY)

Building pad grading (SY)

Dirt disposal (CY)

BF = Board Foot, CY = Cubic Yard, EA = Each, LB = Pound, LF = Linear Foot,
L/S = Lump Sum, SF = Square Foot, SY = Square Yard, SQ = Square (100 square feet)

QUANTITY TAKE-OFF CHECKLISTS (cont.)

Site Work (cont.)

Earthwork (cont.)

Testing (L/S)

Jacking, boring, and piling (LF)

Support and protection (L/S)

Silt fence or turbidity barrier (LF)

Paving

Asphalt paving (SY), including sub-base, base and
  asphalt topping

Curbs (LF), listing each type

Concrete driveway apron, handicapped ramp or
  sidewalk (SF)

Signage (EA) and Stripping (EA or LF)

Storm Drainage

Excavation and backfill (CY)

Manhole demolition (EA)

Pipe demolition (LF)

Manholes (EA), listing different types

Catch basins (EA)

Culverts (LF)

Pipes (LF), listing different materials and dimensions
  (e.g. 18" vs. 30", RCP vs. PVC)

Tie-ins (EA)

Roof drain connection (EA)

Foundation drain connection (EA)

Sump pumps (EA)

BF = Board Foot, CY = Cubic Yard, EA = Each, LB = Pound, LF = Linear Foot,
L/S = Lump Sum, SF = Square Foot, SY = Square Yard, SQ = Square (100 square feet)

QUANTITY TAKE-OFF CHECKLISTS *(cont.)*

Site Work *(cont.)*

Sanitary Sewer

Excavation and backfill (CY)

Open cut and repair (SY)

Manholes (EA), listing different types

Cleanouts (EA)

Pipes (LF), listing different materials and dimensions
   (e.g. 4" vs. 6", DIP vs. PVC)

Septic tanks (EA)

Grease traps (EA)

Lift stations (EA)

Water

Excavation and backfill (CY)

Open cut and repair (SY)

Pipes (LF), listing different materials and dimensions
   (e.g. 2" vs. 3", Copper vs. PVC)

Fittings (EA), including tees, valves, etc.

Backflow preventer (EA)

Test and balance (L/S)

Fire Underground

Excavation and backfill (CY)

Open cut and repair (SY)

Pipes (LF), listing different materials and dimensions
   (e.g. 4" vs. 6", DIP vs. PVC)

Fittings (EA), including tees, valves, etc.

FDC, i.e. fire department connection (EA)

Fire Hydrants (EA)

BF = Board Foot, CY = Cubic Yard, EA = Each, LB = Pound, LF = Linear Foot,
L/S = Lump Sum, SF = Square Foot, SY = Square Yard, SQ = Square (100 square feet)

QUANTITY TAKE-OFF CHECKLISTS (cont.)

Landscaping

Trees (EA)

Shrubs (EA)

Sod (SF)

Seed (SF)

Mulch (SF)

Top soil (CY)

Fertilizing (SF)

Maintenance (L/S)

Irrigation

Sleeves (LF)

Pipes (LF)

Fittings (EA)

Sprinkler heads (EA)

Planter drain (LF)

Interior Selective Demolition

Demolish concrete foundation, columns, beams, and staircases (CY)

Demolish floors, walls, ceilings, and roofs (SF)

Demolish structural steel columns, beams, and joists (EA)

Demolish doors, windows, millwork, and specialty items (EA)

Cutting and patching (L/S)

Temporary fencing (LF)

Temporary partitions (SF)

Shoring and engineering (L/S)

BF = Board Foot, CY = Cubic Yard, EA = Each, LB = Pound, LF = Linear Foot,
L/S = Lump Sum, SF = Square Foot, SY = Square Yard, SQ = Square (100 square feet)

QUANTITY TAKE-OFF CHECKLISTS (*cont.*)

Concrete

Foundation: Including Continuous footing or Grade beams (LF), Spread footing or Pile caps (EA), Thickened edge (LF) etc.

Layout (L/S)

Excavation (CY)

Formwork (SF)

Concrete (CY),

Rebar (LB or TON)

Embeds (EA)

Backfill (CY)

Slabs (SF)

Fine Grading (SF)

Gravel base (SF)

Vapor barrier (SF)

Formwork (SF)

Concrete (CY)

Rebar (LB or TON)

Wire mesh (SF or Roll)

Curing sealing compound (SF, Gallon or Pail)

Finishing and Curing (SF)

Saw cut control joints (LF)

Hardener (SF, Gallon or Pail)

Insulation (SF or Roll)

Support for suspended slab (L/S)

BF = Board Foot, CY = Cubic Yard, EA = Each, LB = Pound, LF = Linear Foot, L/S = Lump Sum, SF = Square Foot, SY = Square Yard, SQ = Square (100 square feet)

QUANTITY TAKE-OFF CHECKLISTS (*cont.*)

Concrete (*cont.*)

Curbs (LF)
Formwork (SF)
Concrete (CY)
Rebar (LB or TON)
Finishing (SF)

Beams (LF), Columns (EA) and Pedestals (EA)
Layout (L/S)
Scaffolding (SF)
Formwork (SF)
Concrete (CY)
Rebar (LB or TON)
Finishing (SF)
Embeds (EA)

Stairs and Landing (EA)
Formwork (SF)
Concrete (CY)
Rebar (LB or TON)
Finishing (SF)
Support (L/S)

Walls (LF or SF)
Formwork (SF)
Concrete (SF)
Rebar (SF)
Insulation (SF)
Dampproofing (SF)

BF = Board Foot, CY = Cubic Yard, EA = Each, LB = Pound, LF = Linear Foot,
L/S = Lump Sum, SF = Square Foot, SY = Square Yard, SQ = Square (100 square feet)

QUANTITY TAKE-OFF CHECKLISTS (*cont.*)

Masonry

Blocks (EA), by type and grade
Bricks (EA)
Stone (SF)
Rebar (LB or TON)
Cell fill concrete (CY)
Mortar (Bags)
Sand (CY or Ton)
Scaffolding (SF)
Control joint filler (LF)
Joint reinforcement (LF)
Flashing (LF)
Wall tie (EA)
Weep hole (EA)
Installing door frames (EA)
Pre-cast lintels and sills (LF or EA)
Wall cleaning (SF)

Metals

Structural Steel

Joist (LF converted to TON)
Decking (SF)
Shapes (LF converted to TON) including beams,
   columns, plates, channels, angles, etc.
Waste and connection (L/S)
Ladders (EA)
Gates (EA)
Stairs (EA)
Panels (SF)
Rails (LF)

BF = Board Foot, CY = Cubic Yard, EA = Each, LB = Pound, LF = Linear Foot,
L/S = Lump Sum, SF = Square Foot, SY = Square Yard, SQ = Square (100 square feet)

QUANTITY TAKE-OFF CHECKLISTS (*cont.*)

Metals (*cont.*)

Ornamental Metals

Stairs (EA)

Rails (LF)

Panels (SF)

Gates (EA)

Trims (LF)

Rough Carpentry

Floors

Girders, Sills, Joists, Trimmers, Headers
  (LF converted to BF)

Bridging (EA)

Sheathing (SF converted to sheet)

Decking (SF)

Walls

Plates, Studs, Headers (LF converted to BF)

Sheathing (SF converted to sheet)

Insulation (SF)

Ceilings

Joists, Trimmers, Headers (LF converted to BF)

Roof

Rafters, Trimmers, Headers, Collar Ties, Ridge and
  Trims (LF converted to BF)

Trusses (EA)

Laminated beams and arches (EA or LF)

Roof sheathing (SF converted to sheet)

BF = Board Foot, CY = Cubic Yard, EA = Each, LB = Pound, LF = Linear Foot,
L/S = Lump Sum, SF = Square Foot, SY = Square Yard, SQ = Square (100 square feet)

QUANTITY TAKE-OFF CHECKLISTS (*cont.*)

Finish Carpentry

    Cabinets (LF)

    Countertops (LF)

    Shelving (LF)

    Trims (LF)

    Stairs (EA)

    Railing (LF)

    Panels (SF)

Roofing

    Bituminous sheets, Shingles, Tile, or Metal
      (SF converted to SQ)

    Felts, sheathing paper, and underlayment (SF)

    Insulation (SF)

    Vapor barrier (SF)

    Down sprouts and gutters (LF)

    Flashing (LF)

    Cant strips (LF)

    Roof pavers (SF)

    Gravel stops (LF)

    Coping (LF)

    Ridge strips (LF)

    Fascias (LF)

    Reglets (LF)

    Sidings (SF)

    Roof panels (SF)

    Skylights (EA)

    Roof hatches, vents and scuppers (EA)

BF = Board Foot, CY = Cubic Yard, EA = Each, LB = Pound, LF = Linear Foot,
L/S = Lump Sum, SF = Square Foot, SY = Square Yard, SQ = Square (100 square feet)

QUANTITY TAKE-OFF CHECKLISTS (cont.)

Doors

Metal doors (EA), including steel, aluminum, bronze, or stainless steel

Wood doors (EA)

Special doors (EA)

Folding doors

Grilles

Overhead doors

Vertical lift doors

Acoustical doors

Automatic entrance doors

Revolving doors

Traffic doors

Frames (EA)

Hardware (EA)

Installation (EA)

Glass

Storefront (SF)

Curtain wall (SF)

Glass doors (EA), including shower doors, sliding or patio doors, etc.

Windows (EA)

Glazing (SF)

Hardware (EA)

Engineering (L/S)

Installation (L/S)

BF = Board Foot, CY = Cubic Yard, EA = Each, LB = Pound, LF = Linear Foot, L/S = Lump Sum, SF = Square Foot, SY = Square Yard, SQ = Square (100 square feet)

QUANTITY TAKE-OFF CHECKLISTS (*cont.*)

Drywall

Stud partition, wood or metal (SF or LF)

Furring, wood or metal (LF)

Gypsum board (SF)

Densglass (SF)

Insulation (SF)

Corner bead (LF)

Mud (Pound, Gallon, or Bucket)

Tape (LF or Box)

Screws and Nails (EA or Box)

Scaffolding (SF)

Cut and patch (L/S)

Finishing (SF)

EIFS/Stucco

Lath, gypsum or metal (SY)

Plaster (SY, CF or sacks)

EIFS (SF)

Trims, moldings and shapes (LF or EA)

BF = Board Foot, CY = Cubic Yard, EA = Each, LB = Pound, LF = Linear Foot,
L/S = Lump Sum, SF = Square Foot, SY = Square Yard, SQ = Square (100 square feet)

QUANTITY TAKE-OFF CHECKLISTS (*cont.*)

Flooring

Carpet, including wall carpet (SF or SY)

Vinyl composition tile (SF or SY)

Terrazzo (SF or SY)

Ceramic tile, including wall tile (SF or SY)

Quarry tile (SF or SY)

Marble tile (SF or SY)

Rubber tile (SF or SY)

Marble (SF or SY)

Brick paver (SF or SY)

Wood flooring (SF or SY)

Stone flooring (SF or SY)

Bamboo flooring (SF or SY)

Base (LF or Yard)

Threshold (LF)

Metal strips (LF)

Acoustical Ceilings

Ceiling tile (SF)

Insulation (SF)

Suspension system (L/S)

Installing special lighting fixtures (EA)

Metal strips (LF)

Bulkheads (LF)

BF = Board Foot, CY = Cubic Yard, EA = Each, LB = Pound, LF = Linear Foot,
L/S = Lump Sum, SF = Square Foot, SY = Square Yard, SQ = Square (100 square feet)

QUANTITY TAKE-OFF CHECKLISTS (*cont.*)

Painting

Interior Painting
Walls (SF)
Ceiling (SF)
Floor (SF)
Columns and beams (SF)
Trim (LF)
Stairs (SF)
Doors and windows (EA or SF)
Structural steel priming or painting (TON or SF)
Wall covering (SF)

Exterior Painting
Siding (SF)
Trim (LF)
Doors and windows (EA or SF)
Masonry (SF)
Caulking (LF)

BF = Board Foot, CY = Cubic Yard, EA = Each, LB = Pound, LF = Linear Foot,
L/S = Lump Sum, SF = Square Foot, SY = Square Yard, SQ = Square (100 square feet)

QUANTITY TAKE-OFF CHECKLISTS (*cont.*)

Specialties

Toilet partitions and accessories (EA)

Chalkboards, tackboards, and markerboards (EA)

Mailboxes (EA)

Flagpoles (EA)

Trash receptacles (EA)

Awnings (LF, SF or L/S)

Lockers and benches (EA)

Shutters, grilles, and louvers (EA)

Fireplaces and stoves (EA)

Bicycle racks (EA)

Signage (EA or L/S)

Equipment

Residential Equipment

Clothes dryer (EA)

Clothes washer (EA)

Dishwasher (EA)

Compactor (EA)

Freezer (EA)

Refrigerator (EA)

Range (EA)

Oven (EA)

Disposer (EA)

Range hoods (EA)

Medicine cabinets (EA)

BF = Board Foot, CY = Cubic Yard, EA = Each, LB = Pound, LF = Linear Foot,
L/S = Lump Sum, SF = Square Foot, SY = Square Yard, SQ = Square (100 square feet)

QUANTITY TAKE-OFF CHECKLISTS (cont.)

Equipment (cont.)

Non-residential Equipment

Loading dock equipment (EA)

Food service equipment (EA)

Lab equipment (EA)

Security and vault equipment (EA)

Library equipment (EA)

Playground equipment (EA)

Athletic equipment (EA)

Furnishings

Blinds and shades (EA)

Draperies and curtains (EA)

Church pews (EA)

Bleachers (EA)

Floor mats (EA)

Casework (EA)

Special Construction

Swimming pool, spa, and sauna (L/S)

Vaults (L/S)

Conveying System

Elevators (EA)

Pneumatic tubing system (EA)

Chutes, including laundry, linen, or waste (EA)

Dumbwaiters (EA)

Lifts (EA)

Escalators (EA)

Moving walks (EA)

BF = Board Foot, CY = Cubic Yard, EA = Each, LB = Pound, LF = Linear Foot,
L/S = Lump Sum, SF = Square Foot, SY = Square Yard, SQ = Square (100 square feet)

QUANTITY TAKE-OFF CHECKLISTS (*cont.*)

Plumbing

Equipment

Sewage pumps (EA)

Sump pumps (EA)

Back-flow preventors (EA)

Water heaters (EA)

Storage tanks (EA)

Expansion tanks (EA)

Shut-off valves (EA)

Pressure reducing valves (EA)

Thermostatic mixing valves (EA)

Shut-off valves (EA)

Air-compressors (EA)

Sand/oil interceptors (EA)

Grease traps (EA)

Trench drains (LF)

Floor drains (EA)

Roof drains (EA)

Piping

Rainwater leaders (LF)

Hot water pipe (LF)

Cold water pipe (LF)

Compressed-air pipe (LF)

Natural gas pipe (LF)

Sanitary waste pipe (LF)

Sanitary vent pipe (LF)

Special piping for medical, lab, kitchen
   equipment, etc. (LF)

BF = Board Foot, CY = Cubic Yard, EA = Each, LB = Pound, LF = Linear Foot,
L/S = Lump Sum, SF = Square Foot, SY = Square Yard, SQ = Square (100 square feet)

QUANTITY TAKE-OFF CHECKLISTS *(cont.)*

Plumbing *(cont.)*

Piping (cont.)

Sleeves (LF)

Fittings (EA)

Clean-outs (EA)

Shut-off valves (EA)

Insulation (LF or SF)

Connections for equipment such as dishwashers,
  dryers, and washers (EA)

Fixtures

Drinking fountains (EA)

Wash fountains (EA)

Bubblers (EA)

Water closets (EA)

Urinals (EA)

Bath tubs (EA)

Laundry tubs (EA)

Lavatories (EA)

Sinks (EA)

Showers (EA)

Wall hydrants/Hose bibs (EA)

BF = Board Foot, CY = Cubic Yard, EA = Each, LB = Pound, LF = Linear Foot,
L/S = Lump Sum, SF = Square Foot, SY = Square Yard, SQ = Square (100 square feet)

QUANTITY TAKE-OFF CHECKLISTS (*cont.*)

Fire Protection

Back-flow preventors (EA)

Fire pumps (EA)

Siamese connection (EA)

Sprinkler valve station (EA)

Fire hose cabinets and racks (EA)

Sprinkler Standpipes (LF)

Sprinkler Heads (EA)

Fittings (EA)

Valves (EA)

Hangers (L/S)

Supports (L/S)

HVAC

Equipment

Air handling units (EA)

Heat pumps (EA)

Electric heaters (EA)

Chillers (EA)

Cooling towers (EA)

Boilers (EA)

Expansion tanks (EA)

Storage tanks (EA)

Heat exchangers (EA)

Exhaust fans (EA)

Condensate pumps (EA)

Humidification and Dehumidification (L/S)

Dust collectors (EA)

Fume hoods (EA)

BF = Board Foot, CY = Cubic Yard, EA = Each, LB = Pound, LF = Linear Foot,
L/S = Lump Sum, SF = Square Foot, SY = Square Yard, SQ = Square (100 square feet)

QUANTITY TAKE-OFF CHECKLISTS (cont.)

HVAC (cont.)

Ductwork

Supply ducts (LF, SF or Tons)

Return ducts (LF, SF or Tons)

Ductwork Insulation (SF)

Louvers, diffusers, registers, dampers, and grilles (EA)

Fittings (EA)

Valves (EA)

Filters (EA)

Thermostats (EA)

Test and balance (L/S)

Electrical

Service and Distribution

Service set-up (L/S)

Switchboards (EA)

Transformers (EA)

House Panels (EA)

Meter cabinets (EA)

Emergency generator (EA)

UPS (EA)

Motor control centre (EA)

Feeder cables (LF)

Transfer switches (EA)

Grounding (L/S)

BF = Board Foot, CY = Cubic Yard, EA = Each, LB = Pound, LF = Linear Foot,
L/S = Lump Sum, SF = Square Foot, SY = Square Yard, SQ = Square (100 square feet)

Electrical (*cont.*)

Power

Receptacle (EA)

GFI breakers (EA)

Junction boxes (EA)

Clocks (EA)

Bells (EA)

Thermostats (EA)

Emergency shut-off switches (EA)

Power supply for mechanical equipment,
    hand-dryers, overhead doors, elevator (L/S)

Lighting

Fixtures (EA)

Signs (EA)

Lamps (EA)

Conduits (LF)

Wires (LF)

Relay panels (EA)

Switches, dimmers, sensors (EA)

Special systems

Fire alarm (L/S)

Telephone, intercom, cable and data systems (L/S)

Audio / video system (L/S)

Security and CCTV systems (L/S)

BF = Board Foot, CY = Cubic Yard, EA = Each, LB = Pound, LF = Linear Foot,
L/S = Lump Sum, SF = Square Foot, SY = Square Yard, SQ = Square (100 square feet)

CHAPTER 3
Bid Day

A bid day is almost always hectic, much like exam day at school. If you just wait for things to happen, then you don't score well. So get ready, and be organized.

PUT TOGETHER A BID ESTIMATE

With the deadline approaching, you need to set up a bid estimate to organize everything. This bid estimate can be made up of two sheets. One is a bid workup sheet, which is a list of trade components with their detailed costs. The other is a bid recap sheet, which summarizes all direct and indirect costs to get a bid total.

1. Bid workup sheet. To set up a bid workup sheet, review drawings and specs thoroughly to make your list of trade components as complete as possible. Do not just list the general CSI divisions. Be more detailed. Separate components that require different pricing methods or quotations. For example, you might want to separate "glass storefront" from "windows," because some glass subs might exclude windows from their quotes.

A bid workup sheet should have the following basic elements:

- Activity ID (or CSI code)
- Description of project activity
- Quantities
- Units
- Material unit price
- Material subtotal
- Labor unit price
- Labor subtotal
- Subcontractor unit price
- Subcontractor subtotal
- Total activity costs
- Sub's Name

Refer to **Figure 3.1** for a blank bid workup sheet. **Figure 3.2** shows a completed preliminary bid workup sheet. Please note all the prices and subs are "plug-ins" for now, waiting for the updates on the bid day.

It is wise to make reasonable expectations about how you will get a price for each line item. Some items are self-performed, and for these you plug in labor and material unit prices as well as quantities. For other items, you will get quotes from your subcontractors and suppliers. However, even if you expect to receive quotes from subs, a detailed trade estimate is quite helpful. For example, you could have counted doors, measured the area of building sidewalk, stone veneer, the length of window sill, etc. Later you could compare your take-off against subs' quotes to identify any

PUT TOGETHER A BID ESTIMATE (cont.)

discrepancies, or use it as a back-up if you fail to receive any quotes at all.

2. Bid recap sheet. The bid recap sheet is where you reach the total bid price. To get there, follow these simple steps:

1. Add up all the costs in bid workup sheet, and get separate totals for material, labor, and sub-contractor prices.
2. Calculate sales taxes for material total and labor burden for labor total.
3. Add material total, labor total, subcontractor total, sales tax, and labor burden together to get a total direct cost.
4. Figure the indirect costs such as overhead, bond, insurance, financing, permit, and development fees, etc.
5. Finally, add the profit to get a grand total bid price.

Figure 3.3 shows a bid recap sheet with all formulas explained. Sales tax and labor burden have been discussed in **Chapter 2**, so now you will learn how to figure indirect costs and profits.

3. Indirect costs

a. Overhead

Overhead is one of your costs, not profit. You must pay for your overhead. There are two types: home office overhead and jobsite overhead. They are calculated differently.

Home Office Overhead. Home office overhead could not be directly tied to a specific job. You must pay for these costs to remain in business. They include salaries and fringe benefits for office personnel, estimating costs, home office rent, utilities, furniture and supplies, etc. Home office overhead is calculated by applying a percentage to total direct costs. Refer to **Figure 3.4** for more information.

Jobsite Overhead. Jobsite overhead is directly related to your job. Overhead includes such things as the costs to run a jobsite office and the salaries for jobsite personnel. Because overhead could be 20 to 40 percent of the total bid, you need to estimate it with great accuracy. Unlike calculating home office overhead, do not just apply a percentage. Instead, follow these guidelines to make a thorough calculation:

1. Define a list of jobsite overhead items you need to include. Use the checklist in **Figure 3.5** to help you.

2. Decide how long it will take to finish the job.

3. Price each item based on job duration or simply by lump sum. Note that the costs for some items are recurring throughout the job, possibly daily, weekly, monthly, or even yearly.

4. Summarize all the costs to get total jobsite overhead. **Figure 3.6** shows the job overhead costs figured for a 4-month project.

To determine the job duration, you might have to prepare a preliminary project schedule. Some commercial software can help with scheduling, but the software could be expensive. For bidding small jobs, drawing a simple diagram on scratch paper might be good enough. In any case, discuss the duration with your project managers or superintendents to get their opinion.

b. Bonds

Some bids require a bid bond to ensure you will sign the contract if the job is awarded to you. Your surety should provide bid bond free or for a small annual service charge (normally no more than $300), so you do not need to add much money for bid bonds. Check in advance to see if a bid bond is required so that you can get the paperwork done on time if necessary.

Another thing to include in the bid price is performance and payment bonds. A performance bond is your guarantee to the owner that you will finish the job according to contract documents. Payment bond, or labor and material bond, promises all labor and material supplied on the job will be paid for and thus protects the owner from any claims. They are normally made out to 100% of the contract amount.

The expense for payment and performance bond is usually 1% to 3% of the total job cost. It can be calculated by using a rate table provided by your surety. A sample rate table is given in **Figure 3.7** with the associated bond calculation for a job of $2.7 million dollars. A safer way recommended to figure out the bonding costs, however, is to contact your surety company directly to get a price quote. By doing so, you can also make certain your surety is willing to bond the job.

c. Insurance

There are many types of insurances you are required to have. Most of them could be included in other costs. For example, it might be easier to include worker's compensation insurance in your home office overhead, while the liability insurance for your jobsite vehicles could be put under job overhead.

A specific kind of insurance you might need to include in the bid price is builder's risk. It protects the job against direct loss. Sometimes it also covers the temporary structures, sheds, materials, and equipment stored on site. The rate normally runs from $0.30 to $1.05 per $100 of construction contract value, depending on a number of factors such as job location, type of construction, your company history, etc. If your bid requires builder's risk insurance, you must get a quote directly from your insurance company. When you have the quote, make sure the deductible is suitably low and the job is protected from all possible hazards such as fire, lightning, wind, flood, etc.

PUT TOGETHER A BID ESTIMATE (cont.)

d. Financing

You might have to include the interest on construction loans. On some residential jobs, instead of receiving monthly payments, you will only receive three payments, one after the completion of the foundation, another after building drying-in, and the last after substantial completion. You will need to have enough funds to cover one third of the construction period before receiving any reimbursement from owners. Otherwise, determine the dollar amounts you want to borrow and the length of time, and then shop around at local banks to see how much they will charge you for interest.

e. Permits and Development Fees

There are two types of permits for each job. One is the specialty permit for each trade, such as the plumbing permit, electrical permit, etc. It is customary for each sub to include such costs in his proposal to the general contractor. The other type of permit, the general building permit, is for the whole job. Its cost depends on the type of construction and is usually based on a rate per thousand dollars of job cost.

In addition, development fee (or impact fee) is charged by local government to the proposed building. Depending on construction type and job location, it could be calculated based on a rate per square foot of the building area.

PUT TOGETHER A BID ESTIMATE *(cont.)*

The information on permits and development fees can be obtained by a quick phone call to the local city or county building department.

4. Profit. Profit is the money you want to make from the job, and it is normally determined by applying a rate to total costs. The rate could run 20% to 25% for small jobs and 5% to 10% on a large one. Before deciding the rate to be used, it is important to have a clear understanding of what you are up against. Evaluate how much risk you are taking and how much money you could make. Then decide whether it is worth taking a gamble.

If you really want the job and bid too low, chances are you will lose money completing the job. A careful analysis of your competition's bidding behavior might help you come up with a competitive bid.

Shown in **Figure 3.8** is a preliminary bid recap sheet based on the information from the bid workup sheet given earlier in **Figure 3.2**.

BID DAY

1. Bid rehearsal. One or two days before the bid, you need to check how the bid is going so far. Use the question list in **Figure 3.9** to evaluate the situation and decide if any changes are necessary. In order to avoid the unpleasant bid-day surprises, it is better to get problems resolved beforehand.

Doing a bid involves a lot of number-crunching. For best results, it is recommended to set up the bid estimate sheets using a spreadsheet program or a specialized estimating software which could work even more effectively. Whichever you choose, the idea is the same: use a bid workup sheet to plug in prices for each trade, and use a bid recap sheet to automatically get a summary. Check your work to make sure all numbers work out right.

Because computers can perform calculations quickly without making errors, you will save time and have confidence that your price is accurate, even if there are last minute adjustments to be made. Suppose there was an electrical price change two minutes before the bid was due, you would only have time to change the electrical number in the bid workup sheet. If you set up everything in spreadsheet format, that change would automatically be transferred to your grand total on the proposal form, including the adjustments for overhead, profit, etc. For this reason, computerized bid forms are a great asset.

2. Working on the bid. A successful bid is usually the result of team effort. If you have a bid team of three, one person may be responsible for receiving and evaluating sitework, landscaping, mechanical, and electrical trades. Another person may do concrete, masonry, structural steel, carpentry, and roofing, while the third person could pick up the rest of the trades including doors, glass, finishes, other specialties, etc.

A completed bid proposal form is shown in **Figure 3.10**. If the proposal is to be submitted in person, the estimator usually needs to finish a little earlier than usual so that all of the numbers can be communicated to the person who will actually deliver the bid. To reduce stress, follow the work flow organization shown in **Figure 3.11**.

The focus of any bid day is to put together a "right" total number. If the estimate is set up in spreadsheet format, the bid total will be automatically generated by summarizing all the costs you put in. Therefore, your time is largely spent updating the bid workup sheet by frequently taking new prices from your colleagues. Make sure each line item has a price (total activity cost) and a name (the name of subcontractor). If you did not get a quote, use a plug number for you to self-perform. Put down your company name as the subcontractor. If you are using a supplier to provide the material and a sub for the labor, both names should be referred. Refer to **Figure 3.12** and **Figure 3.13** for the updated bid workup sheet and bid recap sheet.

While getting a correct total price is important, other bidding requirements should not be forgotten. Before you turn in the bid, check everything one more time. The list of subcontractors to be submitted may need to be updated, due to a lower quote received at the last minute. Accuracy in every section of your bid is important.

3. Cost breakdown. Due to financial needs, owners frequently require a cost breakdown as well as a lump sum price. For example, if you are bidding two buildings in one bid, the owner might want to know the separate costs for each building. This breakdown, sometimes called a "schedule of values," could help the owner understand the billings once the job starts.

Figure 3.14 shows a simplified version of typical cost breakdown requirements. Cost breakdown is just another way to present your estimate, but sometimes it must be submitted as part of the bid. Because of the limited time available on bid day, cost breakdown has been a long-time headache for many estimators.

The general idea to do a cost breakdown is as simple as "1-2-3":

1. Study the requirements, and make a list of cost breakdown codes. You may need to change your estimate accordingly to make it more workable.
2. In your bid workup sheet, assign one cost breakdown code to each line item. This means you must decide which cost code each item should carry.

3. Summarize your estimate by cost code, by adding up the costs for items with the same cost code.

To reach an actual result, you might have to add many numbers. **Figure 3.15** and **Figure 3.16** are a revised bid workup sheet and bid recap sheet to meet the cost breakdown required above. A cost code has been allocated to each item, such as "Site Construction," "Building Structures," etc. The items on the bid recap sheet, including sales tax, labor burden, indirect costs, profit, etc., are assigned the code of "Other Costs".

To get the cost breakdown result, add items with the same cost code together. For example, add up the costs for "Fire Protection," "Plumbing," "HVAC," and "Electrical" because they all carry the same cost code —"Building Systems." The finished cost breakdown is shown in **Figure 3.17**. Always be sure the total costs after the breakdown are equal to the original total bid price. Do not leave anything out.

Sometimes owners want the bid to be submitted in a format other than CSI divisions, such as foundations, 1st floor, 2nd floor, roof, exterior walls, interior finishes, etc.

Then you will need to revise your estimate again. For example, further breakdown your drywall number into exterior wall, 1st floor interior wall, 2nd floor interior wall, etc. After completing this type of bid, you should still have a total price that matches your original bid.

FIGURE 3.1 — BID WORKUP SHEET EXPLAINED

No.	Activity	QTY	Unit	Labor U/P	Labor Subtotal	Material U/P	Material Subtotal	Subcontractor U/P	Subcontractor Subtotal	Item Subtotal	Subcontractors' Name

The math in bid workup sheet:

Labor subtotal = Quantities x Labor unit cost

Material subtotal = Quantities x Material unit cost

Subcontractor subtotal = Quantities x Subcontractor unit cost

Item Subtotal = Material subtotal + Labor subtotal + Subcontractor subtotal

FIGURE 3.2 — PRELIMINARY BID WORKUP SHEET

No.	Activity	QTY	Unit	Labor U/P	Labor Subtotal	Material U/P	Material Subtotal	Subcontractor U/P	Subcontractor Subtotal	Item Subtotal	Subcontractors' Name
1	Sitework	1	L/S	—	—	—	—	$495,678	$495,678	$495,678	Plug
2	Landscaping	1	L/S	—	—	—	—	$230,000	$230,000	$230,000	Plug
3	Termite Control	25000	SF	—	—	$0.12	$3,000	—	—	$3,000	Plug
4	Building Concrete	1	L/S	—	—	—	—	$150,175	$150,175	$150,175	Plug
5	Building Sidewalk	1000	SF	—	—	—	—	$4.00	$4,000	$4,000	Plug
6	Masonry	1	L/S	—	—	—	—	$164,679	$164,679	$164,679	Plug
7	Stone Veneer	1050	SF	—	—	—	—	$25	$26,250	$26,250	Plug
8	Foam Insulation	1	L/S	—	—	—	—	$8,000	$8,000	$8,000	Plug
9	Structural Steel	1	L/S	—	—	—	—	$135,000	$135,000	$135,000	Plug
10	Aluminum Trellis	600	SF	—	—	—	—	$40.00	$24,000	$24,000	Plug

3-14

11	Rough Carpentry	1	L/S	$25,000	$25,000	$15,000	$15,000	—	—	$40,000	Plug
12	Window sills	200	LF	—	—	—	—	$25.00	$5,000	$5,000	Plug
13	Built-Up Roofing	1	L/S	—	—	—	—	$124,900	$124,900	$124,900	Plug
14	Caulking	1	L/S	—	—	—	—	$4,500	$4,500	$4,500	Plug
15	H.M. Doors	3	EA	$125.00	$375	$500.00	$1,500	—	—	$1,875	Plug
16	Wood Doors	2	EA	$90.00	$180	$400.00	$800	—	—	$980	Plug
17	Hardware	5	EA	$50.00	$250	$200.00	$1,000	—	—	$1,250	Plug
18	Glass and Glazing	1	L/S	—	—	—	—	$56,000	$56,000	$56,000	Plug
19	Windows	700	SF	—	—	—	—	$12.00	$8,400	$8,400	Plug
20	Stucco	1	L/S	—	—	—	—	$54,000	$54,000	$54,000	Plug
21	Drywall	1	L/S	—	—	—	—	$120,000	$120,000	$120,000	Plug
22	Acoustical Ceilings	1	L/S	—	—	—	—	$16,000	$16,000	$16,000	Plug

FIGURE 3.2 — PRELIMINARY BID WORKUP SHEET (cont.)

No.	Activity	QTY	Unit	Labor U/P	Labor Subtotal	Material U/P	Material Subtotal	Subcontractor U/P	Subcontractor Subtotal	Item Subtotal	Subcontractors' Name
23	VCT	240	SF	—	—	—	—	$5.00	$1,200	$1,200	Plug
24	Painting	1	L/S	—	—	—	—	$36,500	$36,500	$36,500	Plug
25	Toilet Accessories	1	L/S	$2,000	$2,000	$5,500	$5,500	—	—	$7,500	Plug
26	Awnings	1	L/S	—	—	—	—	$7,000	$7,000	$7,000	Plug
27	Fire Protection	1	L/S	—	—	—	—	$30,000	$30,000	$30,000	Plug
28	Plumbing	1	L/S	—	—	—	—	$8,000	$8,000	$8,000	Plug
29	HVAC	1	L/S	—	—	—	—	$65,000	$65,000	$65,000	Plug
30	Electrical	1	L/S	—	—	—	—	$78,000	$78,000	$78,000	Plug
31	Site lighting	1	L/S	—	—	—	—	$12,000	$12,000	$12,000	Plug

FIGURE 3.3 — BID RECAP SHEET EXPLAINED		
Description	Rate	Amount
Labor Total		
Material Total		
Subcontractor Total		
Total Item Costs		
Add Material Sales Tax	%	
Add Labor Burden	%	
Total Direct Costs		
Indirect Costs		
Overhead		
Bond		
Insurance		
Financing		
Permit		
Development Fees		
Total Indirect Costs		
Total Costs		
Profit	%	
Bid Total		

FIGURE 3.3 — BID RECAP SHEET EXPLAINED *(cont)*.

The math in bid recap sheet:

Labor Total = The sum of "Labor Subtotal" column on bid workup sheet

Material Total = The sum of "Material Subtotal" column on bid workup sheet

Subcontractor Total = The sum of "Subcontractor Subtotal" column on bid workup sheet

Material Sales Tax = Material Total × Sales Tax Rate

Labor Burden = Labor Total x Labor Burden Rate

Total Direct Costs = Labor Total + Material Total
+ Subcontractor Total
+ Material Sales Tax
+ Labor Burden

Total Indirect Cost = Overhead + Bond + Insurance
+ Financing + Permit
+ Development Fees

Profit = (Total Direct Cost + Total Indirect Cost)
× Profit Rate

Bid Total = Total Direct Cost + Total Indirect Cost
+ Profit

FIGURE 3.4 — HOME OVERHEAD COST CALCULATION

Formula:

Rate = Home office overhead last year/
          Construction volume last year

Home office overhead for current bid =
  Rate $\times$ Total direct costs for current bid

Example:

Home office overhead last year:
  $300,000

Construction volume last year:
  $5 million

Home office overhead rate:
  $300,000/$5,000,000 = 6%

Total direct costs for your current bid:
  $1,900,000

Home office overhead for current bid:
  $1,900,000 $\times$ 6% = $114,000

FIGURE 3.5 — JOBSITE OVERHEAD ITEMS

- Job mobilization

- Salaries for supervisory personnel such as project managers and superintendents

- Travel expenses for field personnel

- Jobsite signs

- Trailer hook-up including temporary power

- Trailer monthly rental

- Field office furniture and supplies

- Monthly utilities such as heat, electricity, gas, water, telephone

- Small tools and tool sheds

- Material storage

- Temporary partitions, enclosures, fencing, gates, and barricades

- Portable toilet rentals

- Drinking water

- Safety equipment and first aid

- Drawing reproduction

- Postage for project documents

FIGURE 3.5 — JOBSITE OVERHEAD ITEMS *(cont.)*

- Jobsite vehicles and fuels

- Parking for construction crew

- Scaffoldings

- Hoisting equipment

- Permits and licenses

- Surveying and layout

- Material testing costs for both site and building

- Miscellaneous cutting and patching

- General jobsite daily clean-up

- Final clean-up on completion including glass cleaning

- Monthly progress photos

- Allowance to correct punch-list items

- As-built drawings

- Operation and maintenance

FIGURE 3.6 — JOBSITE OVERHEAD COST CALCULATION

Item	QTY	Unit	Rate	Subtotal
Mobilization	1	l/s	$ 8,000	$ 8,000
Project manager	18	wk	$ 1,500	$ 27,000
Superintendent	18	wk	$ 800	$ 14,400
Project sign	1	l/s	$ 3,000	$ 3,000
Trailer rental	4	mon	$ 3,000	$ 12,000
Set-up trailer	1	l/s	$ 2,500	$ 2,500
Field office furniture	1	l/s	$ 1,000	$ 1,000
Field office supplies	4	mon	$ 200	$ 800
Telephone bills	4	mon	$ 150	$ 600
Fax machine	1	l/s	$ 100	$ 100
Computers and IT support	1	l/s	$ 2,500	$ 2,500
Temporary Power	4	mon	$ 4,000	$ 16,000
Water meters	1	l/s	$ 2,000	$ 2,000
Temporary water	4	mon	$ 800	$ 3,200
Portable toilets rental	18	wk	$ 450	$ 8,100
Safety equipment and first aid	1	l/s	$ 1,500	$ 1,500
Drawing reproduction	1	l/s	$ 2,000	$ 2,000
Postage	1	l/s	$ 500	$ 500
Vehicles and fuels	2	each	$ 1,200	$ 2,400
Forklift	18	wk	$ 700	$ 12,600
Small tools	1	l/s	$ 900	$ 900
Perimeter temporary fence	500	LF	$ 3	$ 1,500
Surveying and layout	1	l/s	$ 700	$ 700
Daily clean-up	90	day	$ 40	$ 3,600
Final clean-up	1	l/s	$ 400	$ 400
Punch-list items	1	l/s	$ 4,000	$ 4,000
As-built drawings	1	l/s	$ 1,300	$ 1,300
Total Jobsite Overhead				**$132,600**

FIGURE 3.7 — BOND COST CALCULATION

An example rate table from your surety for
furnishing payment and performance bonds:

Construction Costs	Bond Rate Per Thousand Dollars
First $100,000	$28.50
Next $400,000	$17.10
Next $2,000,000	$11.40
Next $2,500,000	$8.55
Next $2,500,000	$7.98
Over $7,500,000	$7.41

Suppose your job is about $2.7 million dollars and the
bond cost is:

First $100,000:
$100,000/$1000 \times $28.50 = \qquad $2,850

Next $400,000:
$400,000/$1000 \times $17.10 = \qquad $6,840

Next $2,000,000:
$2,000,000/$1000 \times $11.40 = \qquad $22,800

For the balance of construction costs:
($2,700,000 − $2,000,000 − $400,000
− $100,000)/$1,000 \times $8.55 = \qquad $1,710

Total costs for payment and
performance bonds: \qquad $34,200

Please call your surety to verify the costs you calculated.

FIGURE 3.8 — PRELIMINARY BID RECAP SHEET

Description	Rate	Amount
Labor Total		$27,805
Material Total		$26,800
Subcontractor Total		$1,864,282
Total Item Costs		$1,918,887
Add Material Sales Tax	6%	$1,608
Add Labor Burden	25%	$6,951
Total Direct Costs		**$1,927,446**
Indirect Costs		
Home Office Overhead	10%	$192,745
Jobsite Overhead		$132,000
Payment/Performance Bond		$35,000
Builder's Risk Insurance		$11,000
Financing Interest		$60,000
Permit		$15,000
Development Fees		$20,000
Total Indirect Costs		**$465,745**
Total Costs		**$2,393,191**
Profit	8%	$191,455
Bid Grand Total		**$2,584,646**

FIGURE 3.9 — PRE-BID-DAY QUESTION LIST

1. Does your bid workup sheet include everything required in the scope?

2. If you copied the bid workup sheet from some old bids, have you made applicable changes according to the scope of this bid?

3. Is your bid recap sheet taking numbers from the bid workup sheet correctly?

4. Did architects respond to your questions?

5. Have all addenda been issued and received? How did the addenda affect the bid? Have all affected subs and suppliers been notified about these changes?

6. Have you completed detailed estimates for self-performed trades?

7. Do you have enough coverage from qualified subs for important trades?

8. Have your received the quotes for bonds from your surety?

9. Have your received the quotes for builder's risk insurance from your insurance company?

10. Have you found a room in the company where you will be working on the bid?

FIGURE 3.9 — PRE-BID-DAY QUESTION LIST *(cont.)*

11. Have you gathered all drawings and specs with the latest revisions included?

12. Have you prepared a bid proposal form to be submitted? Is it in spreadsheet format? Could it be linked to your bid estimate sheets?

13. Can you get some of the documents to be submitted ready in advance? (Examples include signatures, copy of business license, insurance certificates, etc.)

14. Does everyone in the office know that you are bidding the job on that day? Who will help you with the bid?

15. Have you gathered all the quotes and kept them organized in a folder?

16. Have you double-checked the time, date, and place for the bid to be submitted?

FIGURE 3.10 — BID PROPOSAL FORM

To: ABC Owner

Re: All American School

From: XYZ Contractor

Having carefully examined the drawings and specifications, visited the site, familiarized with bidding requirements and other factors affecting the costs, the undersigned submits proposals to furnish all labor and materials for constructing the above project as follows:

Base Bid (Express in words and figures):
Two Million Three Hundred Seventy-Five Thousand Five Hundred Eighty-Two Only ($2,375,582) Dollars.

Subject to the attached bid clarifications.

Alternate #1: Use ceramic tiles in lieu of VCT, *add $5,000.*

Alternate #2: Delete fire sprinkler system: *deduct $20,000.*

Receipt of Addendum No. 1 and 2 is acknowledged.

I understand that the Owner has the right to reject any or all bids.

Signature:

Title: President

Date: January 1, 2005

FIGURE 3.11 — TYPICAL BID DAY WORK FLOW

Before starting to work on the bid: have a short meeting with your bid team. Review the general scope of work by going through drawings, specs, and addendums; and discuss requirements of cost breakdown, alternates, unit pricing, owner's allowances, etc.

One hour before the bid is due: begin to take trade prices from your team members and feed them into the bid workup sheet. You might discover some problems in getting prices for certain trades. Ask your colleague to keep working on them until they are resolved.

One half hour before the bid is due: make sure all prices are in the bid workup sheet. Discuss with your team member to resolve any problems immediately.

Ten minutes before the bid is due: check the bid workup sheet and the bid recap sheet one more time, and make sure the numbers match. Transfer the bottom line number to the bid proposal. Make sure all the blanks on the proposal form have been filled out correctly.

Five minutes before the bid is due: get approval from your supervisor or manager and submit the bid proposal form with all other required documents.

After the bid has been submitted: phone the owner or the architect to make sure the bid proposal has been received.

FIGURE 3.12 — UPDATED BID WORKUP SHEET

No.	Activity	QTY	Unit	Labor U/P	Labor Subtotal	Material U/P	Material Subtotal	Subcontractor U/P	Subcontractor Subtotal	Item Subtotal	Subcontractors' Name
1	Sitework	1	L/S	—	—	—	—	$385,000	$385,000	$385,000	ABC Site
2	Landscaping	1	L/S	—	—	—	—	$180,000	$180,000	$180,000	XYZ Land Lease
3	Termite Control	25000	SF	—	—	$0.08	$2,000	—	—	$2,000	ACME Termite
4	Building Concrete	1	L/S	—	—	—	—	$130,500	$130,500	$130,500	Adam Concrete
5	Building Sidewalk	1000	SF	—	—	—	—	$4.00	$4,000	$4,000	Adam Concrete
6	Masonry	1	L/S	—	—	—	—	$175,000	$175,000	$175,000	Block Works
7	Stone Veneer	1050	SF	—	—	—	—	$25.00	$26,250	$26,250	Block Works
8	Foam Insulation	1	L/S	—	—	—	—	$7,500	$7,500	$7,500	Northern Foam
9	Structural Steel	1	L/S	—	—	—	—	$145,000	$145,000	$145,000	Northwest Steel
10	Aluminum Trellis	1	L/S	—	—	—	—	$30,000	$30,000	$30,000	Dilworth Metals

FIGURE 3.12 — UPDATED BID WORKUP SHEET (cont.)

No.	Activity	QTY	Unit	Labor U/P	Labor Subtotal	Material U/P	Material Subtotal	Subcontractor U/P	Subcontractor Subtotal	Item Subtotal	Subcontractors' Name
11	Rough Carpentry	1	L/S	$25,000	$25,000	$15,000	$15,000	—	—	$40,000	General Contractor
12	Window sills	1	L/S	—	—	—	—	$4,500.00	$4,500	$4,500	Creative Millwork
13	Built-Up Roofing	1	L/S	—	—	—	—	$135,000	$135,000	$135,000	Countrywide Roof
14	Caulking	1	L/S	—	—	—	—	$5,600	$5,600	$5,600	Caulking-All
15	H.M. Doors	3	EA	$125.00	$375	$650.00	$1,950	—	—	$2,325	Tom's Doors
16	Wood Doors	2	EA	$90.00	$180	$500.00	$1,000	—	—	$1,180	Tom's Doors
17	Hardware	5	EA	$50.00	$250	$300.00	$1,500	—	—	$1,750	Tom's Doors
18	Glass and Glazing	1	L/S	—	—	—	—	$52,000	$52,000	$52,000	Jerry's Glass
19	Windows	1	L/S	—	—	—	—	$7,560	$7,560	$7,560	Perfect Windows
20	Stucco	1	L/S	—	—	—	—	$65,000	$65,000	$65,000	Stucco Man

No.	Description	Qty	Unit								Vendor
21	Drywall	1	L/S	—	—	—	—	$115,000	$115,000	$115,000	First-class Drywall
22	Acoustical Ceilings	1	L/S	—	—	—	—	$14,530	$14,530	$14,530	South Acoustics
23	VCT	1	L/S	—	—	—	—	$1,250	$1,250	$1,250	All Flooring
24	Painting	1	L/S	—	—	—	—	$38,000	$38,000	$38,000	Complete Painter
25	Toilet Accessories	1	L/S	$2,000	$2,000	$6,000	$6,000	—	—	$8,000	Pacific Supply
26	Awnings	1	L/S	—	—	—	—	$10,000	$10,000	$10,000	Sunroof Awning
27	Fire Protection	1	L/S	—	—	—	—	$28,000	$28,000	$28,000	Northern Fire
28	Plumbing	1	L/S	—	—	—	—	$9,000	$9,000	$9,000	Pandas Piping
29	HVAC	1	L/S	—	—	—	—	$74,580	$74,580	$74,580	Bernard Air
30	Electrical	1	L/S	—	—	—	—	$99,000	$99,000	$99,000	Gordon Electrical
31	Site lighting	1	L/S	—	—	—	—	$13,570	$13,570	$13,570	Richter Electrical

FIGURE 3.13 — UPDATED BID RECAP SHEET		
Description	**Rate**	**Amount**
Labor Total		$27,805
Material Total		$27,450
Subcontractor Total		$1,755,840
Total Item Costs		$1,811,095
Add Material Sales Tax	6%	$1,647
Add Labor Burden	25%	$6,951
Total Direct Costs		**$1,819,693**
Indirect Costs		
Home Office Overhead	10%	$181,969
Job Overhead		$135,000
Payment/Performance Bond		$40,000
Builder's Risk Insurance		$15,000
Financing Interest		$75,000
Permit		$15,000
Impact Fee		$20,000
Total Indirect Costs		**$481,969**
Total Costs		**$2,301,663**
Profit	8%	$184,133
Bid Grand Total		**$2,485,796**

FIGURE 3.14 — COST BREAKDOWN REQUIREMENTS

Bidding Contractors:

Please provide the cost breakdown for the
following items:
(Express in words and figures):

Site Construction:_____

($_____) Dollars

Building Structures: _____

($_____) Dollars

Building Finishes: _____

($_____) Dollars

Building Systems: _____

($_____) Dollars

Other Costs _____

($_____) Dollars

Total Bid Price: _____

($_____) Dollars

FIGURE 3.15 — BID WORKUP SHEET FOR COST BREAKDOWN

No.	Activity	QTY	Unit	Labor U/P	Labor Subtotal	Material U/P	Material Subtotal	Subcontractor U/P	Subcontractor Subtotal	Item Subtotal	Cost Code
1	Sitework	1	L/S	—	—	—	—	$385,000	$385,000	$385,000	Site Construction
2	Landscaping	1	L/S	—	—	—	—	$180,000	$180,000	$180,000	Site Construction
3	Termite Control	25000	SF	—	—	$0.08	$2,000	—	—	$2,000	Site Construction
4	Building Concrete	1	L/S	—	—	—	—	$130,500	$130,500	$130,500	Building Structures
5	Building Sidewalk	1000	SF	—	—	—	—	$4.00	$4,000	$4,000	Building Structures
6	Masonry	1	L/S	—	—	—	—	$175,000	$175,000	$175,000	Building Structures
7	Stone Veneer	1050	SF	—	—	—	—	$25	$26,250	$26,250	Building Structures
8	Foam Insulation	1	L/S	—	—	—	—	$7,500	$7,500	$7,500	Building Structures
9	Structural Steel	1	L/S	—	—	—	—	$145,000	$145,000	$145,000	Building Structures
10	Aluminum Trellis	1	L/S	—	—	—	—	$30,000.00	$30,000	$30,000	Building Structures

#	Description	Qty	Unit							Amount	Category
11	Rough Carpentry	1	L/S	—	$25,000	$15,000	$15,000	—	—	$40,000	Building Finishes
12	Window sills	1	L/S	—	—	—	—	$4,500.00	$4,500	$4,500	Building Structures
13	Built-Up Roofing	1	L/S	—	—	—	—	$135,000	$135,000	$135,000	Building Structures
14	Caulking	1	L/S	—	—	—	—	$5,600	$5,600	$5,600	Building Structures
15	H.M. Doors	3	EA	$125.00	$375	$650.00	$1,950	—	—	$2,325	Building Finishes
16	Wood Doors	2	EA	$90.00	$180	$500.00	$1,000	—	—	$1,180	Building Finishes
17	Hardware	5	EA	$50.00	$250	$300.00	$1,500	—	—	$1,750	Building Finishes
18	Glass and Glazing	1	L/S	—	—	—	—	$52,000	$52,000	$52,000	Building Finishes
19	Windows	1	L/S	—	—	—	—	$7,560.00	$7,560	$7,560	Building Finishes
20	Stucco	1	L/S	—	—	—	—	$65,000	$65,000	$65,000	Building Finishes
21	Drywall	1	L/S	—	—	—	—	$115,000	$115,000	$115,000	Building Finishes
22	Acoustical Ceilings	1	L/S	—	—	—	—	$14,530	$14,530	$14,530	Building Finishes

FIGURE 3.15 — BID WORKUP SHEET FOR COST BREAKDOWN *(cont.)*

No.	Activity	QTY	Unit	Labor U/P	Labor Subtotal	Material U/P	Material Subtotal	Subcontractor U/P	Subcontractor Subtotal	Item Subtotal	Cost Code
23	VCT	1	L/S	—	—	—	—	$1,250.00	$1,250	$1,250	Building Finishes
24	Painting	1	L/S	—	—	—	—	$38,000	$38,000	$38,000	Building Finishes
25	Toilet Accessories	1	L/S	$2,000	$2,000	$6,000	$6,000	—	—	$8,000	Building Finishes
26	Awnings	1	L/S	—	—	—	—	$10,000	$10,000	$10,000	Building Structures
27	Fire Protection	1	L/S	—	—	—	—	$28,000	$28,000	$28,000	Building Systems
28	Plumbing	1	L/S	—	—	—	—	$9,000	$9,000	$9,000	Building Systems
29	HVAC	1	L/S	—	—	—	—	$74,580	$74,580	$74,580	Building Systems
30	Electrical	1	L/S	—	—	—	—	$99,000	$99,000	$99,000	Building Systems
31	Site lighting	1	L/S	—	—	—	—	$13,570	$13,570	$13,570	Site Construction

FIGURE 3.16 — BID RECAP SHEET FOR COST BREAKDOWN

Description	Rate	Amount	
Labor Total		$27,805	See Bid Workup Sheet
Material Total		$27,450	See Bid Workup Sheet
Subcontractor Total		$1,755,840	See Bid Workup Sheet
Total Item Costs		$1,811,095	See Bid Workup Sheet
Add Material Sales Tax	6%	$1,647	Other Costs
Add Labor Burden	25%	$6,951	Other Costs
Total Direct Costs		**$1,819,693**	**See Above**
Indirect Costs			
Home Office Overhead	10%	$181,969	Other Costs
Job Overhead		$135,000	Other Costs
Payment/Performance Bond		$40,000	Other Costs
Builder's Risk Insurance		$15,000	Other Costs
Financing Interest		$75,000	Other Costs
Permit		$15,000	Other Costs
Impact Fee		$20,000	Other Costs
Total Indirect Costs		**$481,969**	**See Above**
Total Costs		**$2,301,663**	**See Above**
Profit	8%	$184,133	Other Costs
Bid Grand Total		**$2,485,796**	**See Above**

FIGURE 3.17 — COST BREAKDOWN RESULTS

Cost Breakdown Summary

Items	Amount
Site Construction	$580,570
Building Structures	$673,350
Building Finishes	$346,595
Building Systems	$210,580
Other Costs	$674,701
Total Bid	**$2,485,796**

CHAPTER 4
Evaluating Quotes

Evaluating price quotes from your subs or suppliers is one of the most important bid day activities. **Figure 4.1** shows a quote from a roofing contractor. Because most quotes are received and evaluated within one day, you need to use your time effectively and do things right.

You should always try to get quotes in writing. No one has a perfect memory, and people tend to forget as time passes. Although quotes could be delivered in person, through mail, or by fax, sometimes when time is running out, taking a number over the phone is your only choice. If so, use a telephone quotation form shown in **Figure 4.2**.

A good way to evaluate quotes is to compare them in standard bid tabulation sheets, or "bid tabs." **Figure 4.3** is an example of such sheets. Refer to the question list in **Figure 4.4** when you read each quote.

In order to compare quotes "apples to apples," you need to first define a list of adjustment factors, such as what should be included in everyone's quote and what should not. These factors might come from your own familiarity with the bid or simply by studying one quote that appears to be complete. Then you can adjust each quote to finalize prices.

EXAMPLE #1 — USE YOUR OWN YARDSTICK

When evaluating prices on an important trade, it is a good idea to measure up sub's quotes with your own "plug" price. That means whenever possible, you should prepare a detailed estimate for each trade, whether you subcontract it or not.

Figure 4.5 shows a bid tab on site construction. In this example, you use your own estimate and compare all quotes by different sections: earthwork (including demo), paving (including curbs, signage, and stripes), utilities (storm drainage, sanitary sewer, water, fire, reuse water, gas, electrical conduits, etc), surveying and layout, mobilization, general conditions, profit, etc. In this example, Sub "B" has a low bid which could be a bidding advantage for you.

For adjustment factors, the basic rule is: *if an item is covered elsewhere by one trade, then take it out from all other trades. You need to make sure only one trade has each item, and that there is no overlapping and no gaps.* In this example, site concrete sidewalk was shown on the site drawings and the building concrete sub does not have that in his quote, so the site sub needs to have it. Grease traps happened to be included in the plumber's proposal, so you can take them out of the site sub's quote.

EXAMPLE #2 — COORDINATE DIFFERENT TRADES

Problem:

You received drywall quotes from three subs for a small commercial building:

Sub "A" quoted interior drywall – $21,500, exterior drywall – $39,000. The exterior price included furnishing and installing pre-engineered metal trusses. Sub "A" also included interior/exterior wood blocking and plywood.

Sub "B" quoted interior drywall $18,500, exterior drywall $29,000. Sub "B" included installing pre-engineered metal trusses furnished by others. Sub "B" excluded any wood blocking or plywood.

Sub "C" quoted interior drywall $20,000, exterior drywall $25,000. Sub "C" has no trusses, no wood or plywood. He was only willing to add $1,800 extra to put plywood sheathing over trusses furnished by others.

All above prices include furnishing, hanging and finishing drywall and metal framing, insulation, scaffolding, equipment, clean-up, etc.

You also had some quotes from metal trusses manufacturers.

The lowest material price is $12,500, including the shipping and taxes.

You did a detailed rough carpentry estimate. Some items are:

Interior wood blocking and plywood sheathing for metal framing: $500

Interior wainscot wood base: $700

Exterior wood blocking for metal framing: $ 800

Exterior truss plywood sheathing: $ 2,000

Labor only to install trusses: $9,500

EXAMPLE #2 — COORDINATE DIFFERENT TRADES (cont.)

Solution:

To determine the low drywall price, start with a proposal that appears to be more complete (sub "A" in this case) and find out what he is quoting. Then adjust other subs' quote by adding items they did not include to match sub "A". Of course, you could also consider adding more items to sub "A" if necessary.

Figure 4.6 shows a completed drywall bid tab. The lowest sub appears to be sub "A" for the total drywall prices. The combination of using Sub "B" for interior and Sub "A" for exterior seems to be lower, but subs may not want to sign contracts for interior or exterior work alone.

The real purpose is more than just determining a correct drywall price. Because now the drywall sub has metal trusses furnished and installed, you do not have to add the number from the truss manufacturer. You can also consider deducting your rough carpentry price. All these adjustments from trade coordination will make your total bid more competitive. However, only deduct the relevant items. In the above example, do not deduct carpentry items not covered by drywall subs. Remember, no gaps and no overlapping.

EXAMPLE #3 — EVALUATE "COMBO" QUOTES

Problem:

Seven subs are bidding part of the job with drywall, acoustical ceiling, and VCT.

Sub 1 will do the drywall and acoustical ceiling for $52,000

Sub 2 will do the drywall for $34,000

Sub 3 will do the acoustical ceiling and VCT for $20,000

Sub 4 will do the drywall, acoustical ceiling and VCT for $60,000

Sub 5 will do acoustical ceiling for $15,000

Sub 6 will do VCT for $9,000

Sub 7 will do acoustical ceiling for $13,000

EXAMPLE #3 — EVALUATE "COMBO" QUOTES (*cont.*)

Solution:

Quotes from Sub #1, #3, #4 are called "Combo" because these subs perform a variety of trades for one combined price. Normally, you will need to call them back and tell them to send you revised quotes with separate numbers for each trade. This is probably the recommended procedure to handle "combo" quotes.

Not all subs are willing to give separate quotes because they fear losing business. They fear that you may hire them to do flooring, ceiling, or wall only, instead of all three combined. In addition, when in a hurry, subs could make mistakes in separating numbers.

If subs are not willing to give separate quotes, you could best compare their prices by following these steps:

1. Make a matrix including all scopes of work for each sub.

2. For every sub, plug in the lowest price for the trade he missed. In this example, the lowest standalone price for acoustical ceiling is $13,000, $9,000 for VCT, $34,000 for drywall. By doing so, you are making every sub "jack of all trades".

3. Finally, add up all numbers for each sub, including the plug-in prices.

Figure 4.7 shows the result for this example. You will see the lowest option is to use Sub 2 for drywall and Sub 3 for acoustical ceiling and VCT.

EXAMPLE #4 — CORRECT ERRORS

Subs frequently make mistakes in their quotes. Some are just because of simple math errors, while others are from incomplete scopes and careless quantity take-off. The best way to deal with suspected errors is to contact that sub and ask him to make corrections. Ironically, sometimes you might make sure the sub is also sending corrected quotes to your competitors, if he is bidding to them as well.

Many times you won't be able to get revised quotes on time, especially when "mistakes" were caused by the imperfection in plans and specs. For example, if different quantities of plants are shown on the landscaping drawings from the ones indicated on the plant schedule, some subs are quoting according to the plant schedule, while others are counting trees off the drawings.

What you can do is to decide what your pricing is based on (normally go with the conservative side), and adjust sub's quotes accordingly. For example, if you decide there should be 36 Red Maples and the sub quoted 34, you should read his quote and find his price for Red Maples (say, $800/EA). Add $800 \times (36 - 34) = \$1,600$ to his base bid. Once you have done that, spell it out clearly in a list of "bid clarifications" to the owner.

EXAMPLE #5 — DEAL WITH ALTERNATES

In many jobs the owner requests pricing for using alternative methods or substitute materials. The alternates are generally listed on the proposal form, and they may add or deduct from your base bid. Because the owner could award the job by evaluating your base bid plus all alternates, you should not ignore the pricing on alternates. Since you must get some alternates from the subs, it is important to make sure subs understand what the requirements are.

In figuring out alternates:

1. Always think about what needs to be done. Normally, if you take something out of the building, then you need to put something back. For example, for an alternate price to decrease the height of the storefront, you need to include the price for increased wall area.

2. Be careful about "voluntary alternates." Your sub might submit an unrelated alternate price. For example, your lowest concrete sub's base price is $23,000 with an alternate add of $2,000 for using 4000 PSI instead of 3000 PSI concrete for beams and columns. But you checked the drawings and found out 4000 PSI concrete was required. Alternate required for the bid turned out to be demolishing the existing building, which has nothing to do with what he quoted. So you need to use $25,000, not $23,000, for concrete.

EXAMPLE #5 — DEAL WITH ALTERNATES *(cont.)*

3. Have your own estimate for the alternates because subs could be less willing to provide numbers for alternates than for base contracts.

4. Make sure your final number for alternates include your indirect costs such as overhead, profit, etc. If the alternate is an "add," then figure a percentage (normally 10% to 15%) to be added to the quotes from subs. If the alternate is a "deduct," then present their quotes as they are.

Refer to **Figure 4.8** for calculations on two alternates. Alternate #1 is based on the subs' quotes, and alternate #2 is based on your own estimate when you fail to get any quotes. In either case, you need to understand what makes up an alternate.

The fear of adding too little or deducting too much is a constant concern in pricing alternates. For example, sub A's base bid is $10,000 with alternate $5,000 add, and sub B's base bid is $8,000 with alternate $9,000 add. You decided to use Sub B. Then at the last minute there is a new quote from sub C. His base bid is $7,000 with alternate $10,000 add. Who you should use is a question of bidding strategy. The best thing to do is probably to contact all three subs to find out what their prices are based on. At the very least, make sure alternate prices you use in your calculations come from the same sub you used for the base bid. For example, if you decided to use Sub C, use his alternate price as well.

FIGURE 4.1 — A ROOFING QUOTE

Re: All American School Project

To: XYZ Contractors

From: Sigma Roofing

Date: January 1st, 2005

Gentlemen:

We take pleasure and great comfort in submitting our bid to you on the above referenced project. All materials are bid in accordance with the contract documents, including Addendum #1 and #2.

Total Price: $125,000

Exclude: roof hatches

Any questions, please call (222) 333-4445.

FIGURE 4.2 — TELEPHONE QUOTATION FORM

Date: _____

Project Name: _____

Trade: _____

Subcontractor/Supplier Name: _____

Contact Person: _____

Telephone: _____

Fax: _____

Addendum Received: _____

Scope of Work: _____

Proposal Amount: _____

Alternates: _____

Inclusions: _____

Exclusions: _____

Quote Taken By: _____

FIGURE 4.3 — BID TABULATION FORM

Bid Tab

Job:		Date:			
Trade:		By:			
Scope of Work		Sub #1	Sub #2	Sub #3	Sub #4
Base Bid Amount					
Adjustments					
1					
2					
3					
4					
5					
Adjusted Amount					
Notes					

FIGURE 4.4 — QUESTION LIST FOR EVALUATING QUOTES

Did you solicit quotes from this sub? If not, how much do you know about him?

Is the sub also bidding to your competition?

Is the quote in accordance with plans and specs?

Is the sub using specified material?

Did the sub review and acknowledge all addendums that affect this trade?

Is the sub aware of schedule requirements and liquidated damage clause?

What are the exclusions and inclusions?

What is the cost for the sub to bond the job for his scope of work?

Did the sub breakdown his quote in the format as required by the bid?

Did the sub make any math errors in his quote?

FIGURE 4.5 — SITEWORK BID TAB

Description		Your Estimate	Sub A	Sub B	Sub C	Sub D
Base Bid		**$903,473**	**$888,619**	**$818,584**	**$930,818**	**$885,250**
Breakdown						
	Earthwork	$132,314	$94,518	$96,131	$77,977	$53,400
	Paving/Signage/Curb	$505,361	$466,098	$452,254	$438,117	$481,850
	Drainage	$138,440	$157,389	$138,074	$175,551	$167,202
	Sewer	$62,333	$83,480	$30,479	$101,658	$75,798
	Water	$24,539	$25,485	$16,362	$105,810	$85,000
	Fire line	$40,486	$61,650	$70,288	INCL	$22,000
	Survey/As-Builts	INCL	INCL	$11,500	INCL	INCL
	Mobilization/G.C.	INCL	INCL	$3,500	$31,705	INCL
Adjustment						
Add	Demo	INCL (@$8,640)	INCL	INCL	INCL	INCL
Add	Dewatering	$27,500	INCL (@ 30,000)	$27,500	INCL	INCL (@$25,000)
Add	Concrete Sidewalk	INCL (@$23,936)	INCL (@$24, 670)	INCL (@$15,900)	$21,502	$21,502
Add	Soil Testing	Not Included	($14,000)	No	No	No
Deduct	Fire Line	($40,486)	($61,650)	($70,288)	($61,650)	($85,000)
Deduct	Grease Trap	No	($8,800)	No	No	No
Adjusted Bid		$890,487	$804,169	$775,796	$890,670	$821,752

FIGURE 4.6 — DRYWALL BID TAB			
Scope	Sub A	Sub B	Sub C
Interior			
Drywall	$21,500	$18,500	$20,000
Wood/Plywood	INCL	$500	$500
Wainscot	INCL	$700	$700
Interior Total	$21,500	$19,700	$21,200
Exterior			
Drywall	$39,000	$29,000	$25,000
Truss Material	INCL	$12,500	$12,500
Truss Labor	INCL	INCL	$9,500
Plywood On Truss	INCL	$2,000	$1,800
Wood Blocking	INCL	$800	$800
Exterior Total	$39,000	$44,300	$49,600
Total Drywall Price	**$60,500**	**$64,000**	**$70,800**

FIGURE 4.7 — MATRIX FOR COMBO QUOTES				
Sub	**Drywall**	**Ceiling**	**VCT**	**Total**
#1	$52,000	Included	$9,000	$61,000
#2	$34,000	$13,000	$9,000	$56,000
#3	$34,000	$20,000	Included	$54,000
#4	$60,000	Included	Included	$60,000
#5	$34,000	$15,000	$9,000	$58,000
#6	$34,000	$13,000	$9,000	$56,000
#7	$34,000	$13,000	$9,000	$56,000

FIGURE 4.8 — PRICING ALTERNATES

Alternate# 1: Add 100 LF of 6' high screen wall

Trade	Price
Concrete	$8,000
Masonry	$3,000
Precast wall cap	$3,000
Stucco	$3,000
Paint	$1,200
Subtotal	**$18,200**
Add Overhead and Profit (10%)	$1,820
Alternate #1 Add	**$20,020**

Alternate #2: Use VCT in lieu of Ceramic Tile

Base Bid	QTY	Unit	U/P	Subtotal
Floor Tile	350	SF	$12.00	$4,200
Tile Base and Wainscot	950	SF	$15.00	$14,250
Total				**$18,450**
Alternate				
Floor VCT	350	SF	$4.00	$1,400
VCT Base and Wainscot	950	SF	$5.00	$4,750
Total				**$6,150**
Alternate #2 Deduct				**$(12,300)**

TRADE EVALUATION CHECKLIST

Site Work

General

1. Is mobilization cost included? Does the site sub need multiple mobilizations?

2. Does the plan call for certified survey and/or as-built?

3. Is site permit cost included if required?

4. Is site material testing cost included if required?

5. Does a temporary access road need to be built?

Demolition

1. Is there any demolition involved, exterior or interior?

2. Do we demolish existing buildings totally? Are there any hazardous materials to be removed? Is there an environmental report available for the existing building?

3. Is there any building interior selective demolition? How does that coordinate with building trades such as fire protection, HVAC, plumbing, and electrical? Is there any temporary shoring and support required?

4. Is a temporary fence or partition included to protect the demolition?

5. Who will remove the trash?

TRADE EVALUATION CHECKLIST *(cont.)*

Site Work *(cont.)*

Earthwork

1. How many cubic yards of dirt are being moved? Cut or fill?

2. What will you do with the dirt excavated? (Re-use it, pile it on-site, or haul it off the site?)

3. Are any existing trees to be removed or relocated?

4. How long does it take for the site sub to get the building pad ready?

5. Did the sub review the soil report? Is dewatering included in the quote?

6. Are any hazardous materials present? Was this an existing landfill site?

7. Are any retention ponds, new or existing, shown on the plans?

8. Is there any roadwork involved?

9. If over-excavation for building foundation is required, is it covered by the site sub or concrete sub?

TRADE EVALUATION CHECKLIST *(cont.)*

Site Work *(cont.)*

Utilities

1. Break down utilities into smaller portions such as storm drainage, sanitary sewer, water and fire. Are there any gas lines or telephone conduits? What work will be done by utility companies?

2. Is the lift station included in the sewer price? Who is providing the power for the lift station?

3. Are the connections to building plumbing, fire, and electrical systems included?

4. Is the site fire underground system included? Who is licensed to install the fire line? (site sub or fire sprinkler sub)

5. Are water meter fees and installation costs included? Is water system chlorinization included?

Paving

1. Is stripping and signage included?

2. Is paving asphalt or concrete?

3. Is any concrete flatwork included like sidewalks or driveway aprons?

4. Does the quote include traffic signalization as well as maintenance of traffic?

TRADE EVALUATION CHECKLIST *(cont.)*

Site Work *(cont.)*

Miscellaneous

1. Will there be any jacking, boring, and piling to install irrigation sleeves, telephone, electrical conduits, etc?

2. Is there any retaining wall or screen wall? Are those walls made of cast-in-place concrete, pre-cast concrete, or masonry?

3. Are there any monument signs? Any details?

4. Is there any guardrail or chain link fence?

5. Are there any site furnishings such as brick pavers, benches, pavilion, tree grates, etc.?

Landscape and Irrigation

1. Is landscape and irrigation to be included? Is it a specified owner's allowance?

2. Which one is more, the quantities shown on drawings or the quantities shown on plant schedule? (Do a plant count to verify.)

3. Are contractors quoting the right plant? (Different gallons of plants will make a price difference.)

4. Is the top soil for plants included?

5. What are the maintenance requirements for the plants?

TRADE EVALUATION CHECKLIST *(cont.)*

Landscape and Irrigation *(cont.)*

6. Are any existing trees to be relocated, protected, or removed? Will there be any tree grates or guards?

7. Are sod, mulch, and seed included in the quote? Does the quote have enough quantities? (Some sods might not show on plans, such as sods along the road.)

8. Are there any irrigation drawings available? Did the sub quote according to plans, or is the quote based on design-build?

9. Where is the water source? (Watch for irrigation wells or pumps.)

10. Is an irrigation back flow preventer included in the quote?

11. Are there any irrigation sleeves that should be included?

12. Is water meter installation for irrigation included? Are water meter fees included?

Concrete

1. Does the building require auger piling? (If yes, you need a separate quote for that.)

2. Do the plans call for tilt-up panels? (If yes, you need a separate quote for that. Make sure engineering and shop drawing costs are included.)

TRADE EVALUATION CHECKLIST *(cont.)*

Concrete *(cont.)*

3. Do the plans call for plant pre-cast structural concrete? (If yes, you need a separate quote for that.)

4. Do the plans call for lightweight concrete? (If yes, you need a separate quote for that.)

5. Is the concrete contractor furnishing concrete, or is it labor-only?

6. Is the concrete contractor furnishing and installing rebar?

7. Does the quote include concrete beam and columns (vertical work)? Or is it just foundations and slabs (flat work)?

8. Does the quote include exterior building sidewalk and dumpster pad?

9. Does the quote include the concrete needed for plumbing, HVAC, electrical equipment pad, electrical conduits, and plumbing trenches?

10. Who is constructing the site concrete sidewalk, driveway, and foundation for the retaining wall? (the site guy or the concrete guy)

11. Does the quote include fine grading for building slab?

12. Does the quote include soil treatment?

13. Does the quote include miscellaneous material necessary for concrete work, such as wood blocking and insulation for freezer slab, and joint sealers for slab on grade?

TRADE EVALUATION CHECKLIST *(cont.)*

Masonry

1. How many blocks or bricks did the masonry contractor figure? (You can call him to get a rough number to compare with your own estimate.)

2. Does the quote include cell fill concrete (material and labor)?

3. Does the quote include rebar (material and labor)?

4. Does the quote include scaffolding and equipment?

5. Are there any fire resistant requirements for the blocks such as 4-hour resistance?

6. Does the quote include bricks, if any? What should be the correct brick material price, if not owner's allowance?

7. If the quote says pre-cast is included, does it mean it includes architectural pre-cast concrete or structural pre-cast lintels?

8. If cast stone is shown on plans, is it included in the quote?

9. Are there site plans for any dumpster enclosures or site block walls?

10. Does the quote include miscellaneous items such as flashing, precast sills, waterproofing, and installation of embedded metals, hollow metal door, and window frames?

TRADE EVALUATION CHECKLIST *(cont.)*

Structural Steel

1. How many tons of steel are included in their quotes?

2. Does the quote include both fabrication and erection? If they are coming from different companies, do their scopes match?

3. In the material portion of the quote, are taxes and freight costs included?

4. Make sure the following miscellaneous steels are included for both material and labor:

   a. Handrails and guardrails

   b. Stairs

   c. Ladders

   d. Steel lintels

   e. Dumpster gates and posts

   f. Steel bollards

   g. Framing for HVAC units

   h. Supporting angles for interior partitions

   i. Supporting steel for exterior building signages

   j. Rainhoods over exterior doors

5. Check specs and structural notes. Are shop drawings to be signed and sealed by professional engineers? (If yes, there are extra costs for that.)

TRADE EVALUATION CHECKLIST *(cont.)*

Carpentry

1. Are there any fire rated or pressure treated requirements for lumber or plywood?

2. Are there any wood trusses, heavy timber, or wood decking present?

3. Check drywall quotes. Are drywall contractors including some rough carpentry for their scope of work? Adjust prices accordingly.

4. Refer to interior design drawings, if any, for more information on millwork. Are quotes from approved manufacturers listed?

5. Is the wood blocking for installing millwork included?

6. Are the countertops made of solid surface or plastic laminate?

7. Check plans to find out if there are any FRP (fiber reinforced panels), vinyl ceilings, cornice trims, crown moldings, wood bases etc. Do you have quotes for these items?

8. Does the quote include the material and labor for installing doors, hardware, and toilet accessories?

9. Are there any furniture, fixtures, and equipment furnished by the owner and installed by the contractor? Figure rough carpentry material and labor for that.

Thermal and Moisture Protection

Roof

1. How many types of roof system are present? (Flat, metal, tile, shingle, etc.)

2. Are roof contractors quoting the approved system? Check specs.

3. Are they approved installers of the specified system?

4. Is roof insulation included? What R-value does it have?

5. Are gutters and down sprouts stainless or aluminum?

6. How are roof penetrations done? Are roof hatches included in the quote? What about skylights?

7. If keeping an existing roof, does the quote include repairing of the old roof?

8. How many years of warranty are roofers quoting, labor warranty and material warranty? What is required?

9. Sometimes at the rear side of the roof parapet, there are some aluminum rain-lock panels. Are roofers including those?

10. Does the quote include testing costs for the roofing system?

11. Does the quote include walking pads or concrete roof pavers?

Thermal and Moisture Protection *(cont.)*

Miscellaneous

1. Read plans and specs to define scope of caulking. For example, does concrete floor require sealing? What about caulking for expansion joints? Are painters including caulking in their quotes?

2. Is there an elevator pit? Do you have quotes on waterproofing? Exterior wall may also need to be water-sealed.

3. Is fireproofing required? Check both architectural and structural drawings as well as specs for the scope of spray-on fireproofing. (Note fire stopping is different from fireproofing. Trade contractors often include fire stopping for their scope of work but not spray-on fireproofing.)

4. Are drywall and acoustical ceiling subcontractors furnishing batt insulation for their scope of work?

5. Did you find any rigid insulation such as those along the perimeter of the building?

6. Did the mason include foam insulation for his blocks?

TRADE EVALUATION CHECKLIST *(cont.)*

Doors

1. How many doors are included in the quote? Do a door count yourself.

2. Are doors pre-machined and pre-finished?

3. How many types of doors are shown? (Note special doors such as overhead doors, as well as regular steel and wood doors.)

4. Are those doors from specified manufacturers?

5. Did door suppliers include material taxes in their proposals? What about freight?

6. Is hardware package included in the price?

7. Are there any fire resistant requirements for the doors?

8. Who is installing the doors? If there are many, did carpenters include the installation of doors in their quotes?

Glass

1. Is there a list of approved manufacturers for a curtain wall system?

2. Is there a list of approved manufacturers for glass and glazing?

3. Is the glass impact resistant? Are hurricane shutters or storm panels required?

4. Who is providing glass for the automatic doors?

Glass *(cont.)*

5. Does the quote include glass lites on steel doors?

6. Is there any glass inside the building?

7. What about windows? Sometimes you have to find a separate quote just for windows.

8. How big are mirrors, if any? Are they to be included under glass or under specialties?

9. Should glass contractors provide signed and sealed shop drawings?

10. Is glass cleaning included in the quote?

11. Does the quote include caulking for glass, glazing, and aluminum framing?

Drywall

1. Does the quote include both drywall and framing?

2. Does the quote include batt insulation for drywall? Does it include all types of drywall required, such as moisture resistant drywall (green board)?

3. According to the building fire protection requirements, does the interior demising wall stop over the ceiling, or does it go all the way to the bottom of the roof deck?

4. Does the quote include any carpentry items?

Drywall *(cont.)*

5. Does the quote include scaffolding and equipment?

6. If this is a renovation project, does it involve patching or repairing existing drywall?

7. Does the quote include installation of hollow metal doors and window frames?

8. If there's metal framing, are signed and sealed shop drawings required?

9. If the quote includes pre-engineered trusses, are they wood or metal? Make sure both material and labor costs are included. Are taxes and freight included?

EIFS and Stucco

1. Does this quote include both EIFS and Stucco?

2. Does the quote include metal lath to support stucco?

3. Did the subcontractor include any drywall?

4. Does the quote include any special finish items such as Fypon trim, foam shapes, or glass reinforced fiber columns?

5. Does the quote include scaffolding and equipment?

TRADE EVALUATION CHECKLIST *(cont.)*

Paint

1. Did the quote cover both interior painting and exterior painting?

2. Are there any vinyl wall coverings that will be included?

3. Are the doors to be stained or painted? Are there any wood trims to be painted?

4. If this is a renovation project, is the new painting to match the existing building? Is the existing building to be re-painted?

5. Does the quote include scaffolding and equipment?

Flooring

1. How many different types of flooring are in this bid? (Hard Tile, VCT, Carpet, Wood, Quartz, Resinous, etc.) Within each category of flooring, further identify each different type. For example, the ceramic tile is different from porcelain tile and marble tile, though they all could be called "hard tile."

2. Does the quote include floor preparation?

3. Is the base included with flooring (e.g. the vinyl base for VCT)?

TRADE EVALUATION CHECKLIST *(cont.)*

Flooring *(cont.)*

4. If flooring goes on the wall, is it included? For example, the ceramic tile on restroom walls, in showers, and on building exterior faces.

5. Do you have quotes on special flooring such as athletic flooring?

6. If there is a renovation project, is the removal of the existing floor included?

7. Is the floor cleaning cost included? (mopping and waxing of VCT, vacuuming of carpet, etc.)

Ceiling

1. What type of ceiling is required? Sometimes there is no reflected ceiling plan, but the finish schedule might specify the ceiling.

2. Is insulation for the ceiling required?

3. Are there any acoustical panels on the wall?

4. Are cut-outs for sprinkler heads included?

TRADE EVALUATION CHECKLIST *(cont.)*

Specialties

1. Who are the approved manufacturers?

2. Are installation costs included in the quote? What about Freight and taxes?

3. Are all big cost items included such as hand dryers, lockers, and benches?

4. Did you check both site and building plans for fencing, trash receptacles, bike racks, etc.?

5. A list of questions for pavers

   a. What type of pavers? Concrete or brick? 4x4 or 4x8?

   b. How many SF pavers?

   c. Any specific manufacturer?

   d. Are they to be cleaned and sealed?

   e. What is the pave sub-base? Asphalt or concrete?

6. For awnings, are they canvas or metal? If metal, are they steel or aluminum? Do they require signed and sealed drawings?

TRADE EVALUATION CHECKLIST *(cont.)*

Equipment, Furnishings and Special Construction

1. Is the owner to furnish and/or install these items?

2. Does the installation only involve "loose" labor, or require mechanical and electrical connection? Are there any site storage costs included?

3. For a swimming pool, how soon can it get done? Is the pool deck included? Is the sub including the excavation, concrete, water supply, and electrical for the pool?

Elevator

1. How many floors in the building, and how high is each floor? Is an elevator required?

2. Who is the approved elevator manufacturer?

3. What is the loading capacity for the elevator?

4. Is an elevator inspection cost included in the quote?

5. Is the flooring in the elevator cab included?

6. What is the lead time for elevator delivery after approval of shop drawings?

TRADE EVALUATION CHECKLIST *(cont.)*

Plumbing

1. Is a plumbing permit included?

2. Coordinate with site work. Does the plumbing quote cover all the work 5 feet outside the building? Make sure the connections with site water line are covered.

3. Is roof drainage included?

4. Is excavation and backfill for underground work included?

5. Does the quote include all plumbing fixtures as required? Are fixtures quoted from approved manufacturers? Make sure large fixtures are included, such as grease traps, water heaters, sump pumps, etc.

6. If this is a renovation project, does it involve demolishing or modifying the existing plumbing system? Is concrete cutting and patching included?

7. Coordinate with other mechanical work. Make sure items such as condensate drains are covered, either by the HVAC sub or by the plumber.

8. Are costs for as-built drawings included?

9. Are water meters included?

10. Does the quote include concrete equipment pads?

TRADE EVALUATION CHECKLIST *(cont.)*

Fire Protection

1. Is a fire sprinkler system required? Are there any fire sprinkler drawings? Also refer to reflected ceiling sheets. Check local building codes if no information is shown on drawings.

2. Is the quote based on plans and specs, or just design-build?

3. Is a fire sprinkler permit included?

4. Is a fire pump included?

5. Coordinate with site work. Does the sprinkler quote include any site fire underground lines? Does the quote cover all the work 5 feet outside the building? Make sure the connections with site fire line are covered.

6. If the building has exterior canopies, is a fire sprinkler system required in the exterior canopies?

7. Does the quote include costs for shop drawings and as-built drawings?

8. If this is a renovation project, does it involve demolishing or modifying the existing fire sprinkler system?

TRADE EVALUATION CHECKLIST *(cont.)*

HVAC

1. Is an HVAC permit included? Is a refrigeration system included?

2. Are roof top HVAC units to be furnished by the owner? Who will provide curbs and wood blocking for roof top HVAC units? Is hoisting for the units included?

3. Is HVAC equipment quoted from approved manufacturers?

4. Who is furnishing condensate drainage piping from the roof, the plumber or the HVAC contractor?

5. If this is a renovation project, does it involve demolishing or modifying the existing HVAC system?

6. What kind of ductwork is included? Sheet metal or fiberglass?

7. Does the quote include thermostats? Make sure the electrician is including wiring for such devices.

8. Are costs for shop drawings included?

9. Does the quote include certified test and balance?

10. Does the quote include concrete equipment pads?

Electrical

Site Electrical

1. Does the quote include temporary power, including the power to hook-up the GC's trailer? Where is the transformer location?

2. Who will provide the primary connection, i.e. from power company main line to transformer?

3. Who will provide the secondary connection, i.e. from the transformer into the building?

4. Is there site lighting involved in this bid? Are site light poles buried or mounted on the bases? Who will provide the light pole bases as well as the light poles?

5. Are there any site underground telephone or electrical conduits?

6. Who is providing power for site monument signs, if any?

7. Who is providing power for the irrigation system?

8. Are utility company fees included in the quote?

9. Is excavation and backfill included?

Electrical *(cont.)*

Building Electrical

1. Is an electrical permit included?

2. Are there any special systems required such as a fire alarm system? If yes, are they included in the quote?

3. What about visual, audio, and data systems? Sometimes the owner will provide the system himself but not the conduits for these systems.

4. Are lighting fixtures and electrical panels to be supplied by the owner? Are there special requirements for lighting fixtures? Check specs and interior design drawings.

5. Does the quote include the power for mechanical equipment and FFE hook-up?

6. Are there any roof lights? Is the power for exterior building signage included?

7. If this is a renovation project, does it involve demolishing or modifying the existing electrical system? Is there concrete cutting and patching for that?

8. Does the quote include concrete equipment pads?

CHAPTER 5
Post-Bid Review

No matter what a bid turns out to be, it is recommended for you to go through everything you did. This "post-bid review" process is ignored by some estimators because they are often discouraged by the bad bid outcome and decide not to look at the bid again. A review is obviously needed when you are awarded the job because you will build it soon. But by studying the jobs you did not get, valuable lessons could be learned to help improve future chances.

REVIEWING PROCESS

The process for a post-bid review is as simple as "1-2-3":

1. Gather the paperwork. Create a post-bid document folder. See **Figure 5.1** for a list of documents you need to include.

2. Analyze the bid. Check **Figure 5.2** for a list of some common bidding mistakes. Besides mistakes, there should be a few positive things you can learn from. For example, you might have received competitive prices from some new subs.

REVIEWING PROCESS (*cont.*)

Then you can contact them afterwards to find out more about them. If they are qualified to do the work, invite them to bid your next job.

3. Update cost database. Every time you bid a job, there is a lot of information on labor and material prices. By studying the quotes, you could learn roughly how much a solid-core wood door with frame and hardware costs. You could also consider establishing a matrix to compare similar jobs over time. **Figure 5.3** compares the job you just bid against some similar jobs you did in the past. It lists the size and value for each job with detailed trade cost breakdown. If you continually update the matrix, soon you will have a lot of information available to help with future bids.

COORDINATING WITH PROJECT MANAGEMENT

If you are fortunate enough to be chosen for a job, then a project management (PM) team is formed including project manager, field superintendent, and supporting staff. As the estimator who wins the job, you are one of the most valuable resources for the PM team. Remember, now you are working together as a team to control the costs and make the profit.

To start, have a formal coordination meeting with the PM team. Although you are just trying to give the team a general idea about what the job looks like, the communication is very important and the meeting should be more than just handing over a set of plans and specs.

Refer to **Figure 5.4** for a suggested meeting agenda.

After the job starts, you are expected to provide continuous support to the PM team. You will still need to prepare detailed trade estimates, especially when your company self-performs some work. Because your estimate is now meant to be put on the material purchase order, greater accuracy is needed than an estimate for bidding purposes.

Take the following advice.
a. **Resolve problems in documents ASAP.** If there were some problems in the drawings at the time of the bid, now they need to be cleared with the architect to allow an accurate estimate. Also make sure your estimate is based on the current documents if they have been changed from the time of the bid.
b. **Update your price.** Get the latest quotes on material and labor to make sure it is more suitable for self-performing than hiring subs.
c. **Allow reasonable waste.** You do not want to be short. If you are short, the field crew will run out of material and the job will be delayed. Consider the labor productivity and allow reasonable waste in the estimate. Also make the purchase order detailed and clear enough for proper material ordering and delivery.

FIGURE 5.1 — POST-BID DOCUMENT FOLDER

❑ A copy of bid proposal (bid proposal forms, cost breakdown, unit price, alternates, bid clarifications, fax confirmation, etc.)

❑ A copy of the bid estimate, including the bid workup sheet and bid recap sheet

❑ All the quotes from subs organized by bid tabs

❑ Pre-bid checklist

❑ Documents review notes

❑ Site investigation report

❑ Detailed trade estimates you did for self-performed trades

❑ Contract documents (drawings, specs, addendums, etc.)

FIGURE 5.2 — COMMON BIDDING MISTAKES

a. **Not including required items:** scope omission is perhaps the most serious mistake.

b. **Simple math errors:** you may have incorrectly added or subtracted numbers, or used a wrong formula or conversion factor.

c. **Measurement errors:** you could have used the wrong scale for reduced-size drawings. For example, if the drawings were half-size and you used the scale as shown, then your area was reduced by 25%, not the 50% reduction you may have assumed.

d. **Incorrect material or labor unit prices:** insufficient price updates from suppliers.

e. **Underestimated job duration:** you did not have enough money to cover jobsite overhead and risk paying for possible liquidated damages if you were not able to finish on time.

f. **"Buy-out" price cuts:** you intentionally reduced the quotes from subcontractors or suppliers, hoping to increase competitiveness. This will only bring problems later, as subs might not cut their prices. Even if they do, they may question your honesty in doing business and decide not to bid with you in the future.

g. **"Voluntary" price cuts:** You intentionally reduce your overhead or profits to get the job. The eagerness is not a valid excuse. Construction is a business that you should make reasonable profit on.

	Project #1	Project #2	Project #3	Project #4	Today's Bid
FIGURE 5.3 — COST MATRIX FOR SIMILAR JOBS					
SF	9500	10000	8000	8500	9000
Total Price	$565,050	$576,000	$515,050	$534,300	$564,600
Price Per SF	$59	$58	$64	$63	$63
Termite	$200	$500	$400	$400	$600
Concrete	$120,000	$70,000	$70,000	$90,000	$110,000
Masonry	INCL	$40,000	$40,000	$35,000	$36,000
Foam Insulation	$1,400	$1,600	$1,800	$1,500	$2,000
Structural Steel	$45,000	$40,000	$37,000	$31,000	$31,000
Trusses	$5,000	$4,000	$2,500	INCL	INCL
Rough Carpentry	$35,000	$30,000	$30,000	$20,000	$25,000
Finish Carpentry	INCL	$20,000	$18,000	$15,000	$16,000
B.U.R.	$75,000	$70,000	$80,000	$75,000	$76,000
Batt Insulation	$1,500	$2,000	INCL	INCL	INCL
Caulking	INCL	$2,000	$1,000	$1,600	$1,500

Doors and Hardware	$4,000	$3,000	$3,500	$3,500	$4,500
Shutters	$10,000	$8,000	$9,000	$7,500	$8,000
Automatic Doors	$9,000	$8,000	$10,000	$11,000	$10,000
Glass	$15,000	$16,000	$14,000	$17,000	$18,000
Painting	$13,000	$11,000	$15,000	$12,000	$13,000
Drywall	$42,000	$12,000	$12,000	$9,000	$10,000
Stucco	INCL	$32,000	$25,000	$23,000	$28,000
Acoustical Ceiling	$9,000	$4,000	$6,000	$7,000	$8,000
VCT and Carpet	INCL	$10,000	$8,000	$9,000	$10,000
Hard Tile	$24,000	$13,000	$15,000	$13,000	$12,000
Specialties	$950	$900	$850	$800	$1,000
Fire Sprinkler	$5,000	$8,000	$6,000	$7,000	$4,000
HVAC	$40,000	$60,000	$30,000	$45,000	$50,000
Plumbing	$20,000	$25,000	$30,000	$25,000	$20,000
Electrical	$90,000	$85,000	$50,000	$75,000	$70,000

FIGURE 5.4 — ESTIMATING/PM COORDINATION MEETING

1. **Have a job overview.** Everyone should know the job location, building square footage, site acreage, etc.

2. **Review contract documents**, including drawings, specs, addendums, general and special conditions, etc.

3. **Review the bid estimate.** Identify each component that makes up the building. What assumptions did you make in pricing the job? Are they reasonable?

4. **Review the list of subs and suppliers.** Are they reliable to honor their bids? Are they capable of performing the work? Is there going to be a price escalation for some trades?

5. **Review the estimate** for work performed by your own forces. Did you assume realistic labor productivity in preparing your estimate?

6. **Review the preliminary schedule** and the list of jobsite overhead items. Start a formal baseline schedule to control the job progress.

7. **Evaluate risks.** Everyone needs to know what the job budget is, that is the contract amount you finally signed with the owner, not necessarily the proposed bid. Make a list of factors that might put the job over budget or behind schedule.

CHAPTER 6
Technical Reference

Disclaimer

Data tables, including estimating formulas and labor hours, represent the author's best judgment and care for the information published. Instructions for these data tables, when given, should be carefully studied before using them. The numerical results for any data table is affected by numerous specific project factors such as design standards, site conditions, labor productivity, material waste, etc. Neither the author nor the publisher is responsible for any losses or damages with respect to the accuracy, correctness, value and sufficiency of the data, methods, and other information contained herein.

MATH FORMULAS AND TABLES

Areas of Common Geometric Shapes

Shape	Formula
b a	**Parallelogram** Area = a × b
a c b	**Trapezoid** Area = c × 1/2 × (a + b)
a c b hypotenuse (c) = $\sqrt{\left(a^2 + b^2\right)}$	**Right Triangle** Area = 1/2 × a × b
b a	**Regular Triangle** Area = 1/2 × a × b
a b	**Circle** Area = 3.1416 × a^2 Circumference = 3.1416 × b = 6.2832 × a b (Diameter) = 2 × a (Radius)

6-2

MATH FORMULAS AND TABLES (*cont.*)

Volumes of Common Geometric Shapes

Shape	Formula
	Cylinder Volume = $3.1416 \times a/2 \times a/2 \times b$ $= 0.7854 \times a^2 \times b$
	Pyramid Volume = $1/3 \times a \times b \times c$
	Cone Volume = $1/3 \times 3.1416 \times b \times b \times c$ $= 1.0472 \times b^2 \times c$ Or Volume = $0.3518 \times a^2 \times c$
	Sphere Volume = $1/6 \times 3.1416 \times a \times a \times a$ $= 0.5236 \times a^3$

MATH FUNCTIONS FOR NUMBERS 1 TO 100				
Number	Square	Cube	Square Root	Cubic Root
1	1	1	1.0000	1.0000
2	4	8	1.4142	1.2599
3	9	27	1.7321	1.4422
4	16	64	2.0000	1.5874
5	25	125	2.2361	1.7100
6	36	216	2.4495	1.8171
7	49	343	2.6458	1.9129
8	64	512	2.8284	2.0000
9	81	729	3.0000	2.0801
10	100	1000	3.1623	2.1544
11	121	1331	3.3166	2.2240
12	144	1728	3.4641	2.2894
13	169	2197	3.6056	2.3513
14	196	2744	3.7417	2.4101
15	225	3375	3.8730	2.4662
16	256	4096	4.0000	2.5198
17	289	4913	4.1231	2.5713
18	324	5832	4.2426	2.6207
19	361	6859	4.3589	2.6684
20	400	8000	4.4721	2.7144
21	441	9261	4.5826	2.7589
22	484	10648	4.6904	2.8020
23	529	12167	4.7958	2.8439
24	576	13824	4.8990	2.8845
25	625	15625	5.0000	2.9240

MATH FUNCTIONS FOR NUMBERS 1 TO 100 (cont.)

Number	Square	Cube	Square Root	Cubic Root
26	676	17576	5.0990	2.9625
27	729	19683	5.1962	3.0000
28	784	21952	5.2915	3.0366
29	841	24389	5.3852	3.0723
30	900	27000	5.4772	3.1072
31	961	29791	5.5678	3.1414
32	1024	32768	5.6569	3.1748
33	1089	35937	5.7446	3.2075
34	1156	39304	5.8310	3.2396
35	1225	42875	5.9161	3.2711
36	1296	46656	6.0000	3.3019
37	1369	50653	6.0828	3.3322
38	1444	54872	6.1644	3.3620
39	1521	59319	6.2450	3.3912
40	1600	64000	6.3246	3.4200
41	1681	68921	6.4031	3.4482
42	1764	74088	6.4807	3.4760
43	1849	79507	6.5574	3.5034
44	1936	85184	6.6332	3.5303
45	2025	91125	6.7082	3.5569
46	2116	97336	6.7823	3.5830
47	2209	103823	6.8557	3.6088
48	2304	110592	6.9282	3.6342
49	2401	117649	7.0000	3.6593
50	2500	125000	7.0711	3.6840

MATH FUNCTIONS FOR NUMBERS 1 TO 100 (*cont.*)				
Number	Square	Cube	Square Root	Cubic Root
51	2601	132651	7.1414	3.7084
52	2704	140608	7.2111	3.7325
53	2809	148877	7.2801	3.7563
54	2916	157464	7.3485	3.7798
55	3025	166375	7.4162	3.8030
56	3136	175616	7.4833	3.8259
57	3249	185193	7.5498	3.8485
58	3364	195112	7.6158	3.8709
59	3481	205379	7.6811	3.8930
60	3600	216000	7.7460	3.9149
61	3721	226981	7.8102	3.9365
62	3844	238328	7.8740	3.9579
63	3969	250047	7.9373	3.9791
64	4096	262144	8.0000	4.0000
65	4225	274625	8.0623	4.0207
66	4356	287496	8.1240	4.0412
67	4489	300763	8.1854	4.0615
68	4624	314432	8.2462	4.0817
69	4761	328509	8.3066	4.1016
70	4900	343000	8.3666	4.1213
71	5041	357911	8.4261	4.1408
72	5184	373248	8.4853	4.1602
73	5329	389017	8.5440	4.1793
74	5476	405224	8.6023	4.1983
75	5625	421875	8.6603	4.2172

MATH FUNCTIONS FOR NUMBERS 1 TO 100 (*cont.*)

Number	Square	Cube	Square Root	Cubic Root
76	5776	438976	8.7178	4.2358
77	5929	456533	8.7750	4.2543
78	6084	474552	8.8318	4.2727
79	6241	493039	8.8882	4.2908
80	6400	512000	8.9443	4.3089
81	6561	531441	9.0000	4.3267
82	6724	551368	9.0554	4.3445
83	6889	571787	9.1104	4.3621
84	7056	592704	9.1652	4.3795
85	7225	614125	9.2195	4.3968
86	7396	636056	9.2736	4.4140
87	7569	658503	9.3274	4.4310
88	7744	681472	9.3808	4.4480
89	7921	704969	9.4340	4.4647
90	8100	729000	9.4868	4.4814
91	8281	753571	9.5394	4.4979
92	8464	778688	9.5917	4.5144
93	8649	804357	9.6437	4.5307
94	8836	830584	9.6954	4.5468
95	9025	857375	9.7468	4.5629
96	9216	884736	9.7980	4.5789
97	9409	912673	9.8489	4.5947
98	9604	941192	9.8995	4.6104
99	9801	970299	9.9499	4.6261
100	10000	1000000	10.0000	4.6416

CONVERTING IMPERIAL MEASURES TO METRIC MEASURES

From	To	Multiply By
Length		
inch (in)	centimeter (cm)	2.54
inch (in)	meter (m)	0.0254
inch (in)	millimeter (mm)	25.4
foot (ft)	centimeter (cm)	30.48
foot (ft)	meter (m)	0.3048
foot (ft)	millimeter (mm)	304.8
yard (yd)	meter (m)	0.9144
Area		
square foot (sq ft)	square meter (sq m)	0.0929
square inch (sq in)	square centimeter (sq cm)	6.4516
square inch (sq in)	square millimeter (sq mm)	645.163
square yard (sq yd)	square meter (sq m)	0.8361
square (sq)	square meter (sq m)	9.2903
acre (ac)	square meter (sq m)	4047
acre (ac)	hectare (ha)	0.4047
Volume		
cubic inch (cu in)	cubic centimeter (cu cm)	16.3872
cubic foot (cu ft)	cubic centimeter (cu cm)	28317
cubic foot (cu ft)	cubic meter (cu m)	0.0283
cubic foot (cu ft)	liter (l)	28.317
cubic yard (cu yd)	cubic meter (cu m)	0.7646
American gallon (gal)	liter (l)	3.7853
Weight/Density		
pound (lb)	kilogram (kg)	0.4536
short ton, 2000 lb	kilogram (kg)	907.1848
Pound per linear foot (lb/ft)	kilogram per meter (kg/m)	1.488
pound per square inch (psi)	kilopascal (kPa)	6.894
pound per square inch (psi)	megapascal (MPa)	0.0069
pound per cubic foot (pcf)	kilogram per cubic meter (kg/m^3)	16.02

CONVERTING METRIC MEASURES TO IMPERIAL MEASURES

From	To	Multiply By
Length		
centimeter (cm)	inch (in)	0.3937
millimeter (mm)	foot (ft)	0.0033
millimeter (mm)	inch (in)	0.0394
meter (m)	inch (in)	39.37
meter (m)	foot (ft)	3.2808
meter (m)	yard (yd)	1.0936
Area		
square meter (sq m)	square foot (sq ft)	10.7639
square meter (sq m)	square yard (sq yd)	1.196
square centimeter (sq cm)	square inch (sq in)	0.155
square millimeter (sq mm)	square inch (sq in)	0.0016
hectare (ha)	acre (ac)	2.471
hectare (ha)	square foot (sq ft)	107639
Volume		
cubic centimeter (cu cm)	cubic inch (cu in)	0.061
cubic meter (cu m)	cubic foot (cu ft)	35.3145
cubic meter (cu m)	cubic yard (cu yd)	1.308
cubic meter (cu m)	board foot (bf)	423.783
cubic meter (cu m)	American gallon (gal)	264.2
liter (l)	American gallon (gal)	0.2642
Weight/Density		
kilogram (kg)	pound (lb)	2.2046
Metric ton	short ton, 2000 lb	1.1023
kilogram per meter (kg/m)	Pound per linear foot (lb/ft)	0.672
kilopascal (kPa)	pound per square inch (psi)	0.145
megapascal (MPa)	pound per square inch (psi)	145
kilogram per cubic meter (kg/m^3)	pound per cubic foot (pcf)	0.0624

CONVERTING INCHES TO DECIMALS

Inches	Inches in Decimals	Feet In Decimals	Millimeters	Meters
1/16	0.0625	0.0052	1.5875	0.0016
1/8	0.1250	0.0104	3.1750	0.0032
3/16	0.1875	0.0156	4.7625	0.0048
1/4	0.2500	0.0208	6.3500	0.0064
5/16	0.3125	0.0260	7.9375	0.0079
3/8	0.3750	0.0313	9.5250	0.0095
7/16	0.4375	0.0365	11.1125	0.0111
1/2	0.5000	0.0417	12.7000	0.0127
9/16	0.5625	0.0469	14.2875	0.0143
5/8	0.6250	0.0521	15.8750	0.0159
11/16	0.6875	0.0573	17.4625	0.0175
3/4	0.7500	0.0625	19.0500	0.0191
13/16	0.8125	0.0677	20.6375	0.0206
7/8	0.8750	0.0729	22.2250	0.0222
15/16	0.9375	0.0781	23.8125	0.0238
1	1.0000	0.0833	25.4000	0.0254
2	2.0000	0.1667	50.8000	0.0508
3	3.0000	0.2500	76.2000	0.0762
4	4.0000	0.3333	101.6000	0.1016
5	5.0000	0.4167	127.0000	0.1270
6	6.0000	0.5000	152.4000	0.1524
7	7.0000	0.5833	177.8000	0.1778
8	8.0000	0.6667	203.2000	0.2032
9	9.0000	0.7500	228.6000	0.2286
10	10.0000	0.8333	254.0000	0.2540
11	11.0000	0.9167	279.4000	0.2794
12	12.0000	1.0000	304.8000	0.3048

MASTERFORMAT AND UNIFORMAT

OLD MASTERFORMAT 1995 DIVISIONS AND TITLES

Series 0 — Bidding and Contracting Requirements

Division 01 General Requirements

Division 02 Site Work

Division 03 Concrete

Division 04 Masonry

Division 05 Metals

Division 06 Wood, Plastics

Division 07 Thermal and Moisture Protection

Division 08 Doors and Windows

Division 09 Finishes

Division 10 Specialties

Division 11 Equipment

Division 12 Furnishings

Division 13 Special Construction

Division 14 Conveying Equipment

Division 15 Mechanical

Division 16 Electrical

NEW MASTERFORMAT 2004 DIVISIONS AND TITLES

Procurement And Contracting Requirements Group

Division 00 Procurement and Contracting
Requirements

Specifications Group

General Requirements Subgroup

Division 01 General Requirements

Facility Construction Subgroup

Division 02 Existing Conditions

Division 03 Concrete

Division 04 Masonry

Division 05 Metals

Division 06 Wood, Plastics, and Composites

Division 07 Thermal and Moisture Protection

Division 08 Openings

Division 09 Finishes

Division 10 Specialties

Division 11 Equipment

Division 12 Furnishings

Division 13 Special Construction

Division 14 Conveying Equipment

Division 15 Reserved

Division 16 Reserved

Division 17 Reserved

Division 18 Reserved

Division 19 Reserved

MASTERFORMAT AND UNIFORMAT *(cont.)*

Facility Services Subgroup

Division 20 Reserved

Division 21 Fire Suppression

Division 22 Plumbing

Division 23 Heating, Ventilating, and Air Conditioning

Division 24 Reserved

Division 25 Integrated Automation

Division 26 Electrical

Division 27 Communications

Division 28 Electronic Safety and Security

Division 29 Reserved

Site And Infrastructure Subgroup

Division 30 Reserved

Division 31 Earthwork

Division 32 Exterior Improvements

Division 33 Utilities

Division 34 Transportation

Division 35 Waterway and Marine Construction

Division 36 Reserved

Division 37 Reserved

Division 38 Reserved

Division 39 Reserved

MASTERFORMAT AND UNIFORMAT (cont.)

Process Equipment Subgroup

Division 40 Process Integration

Division 41 Material Processing and Handling
Equipment

Division 42 Process Heating, Cooling, and Drying
Equipment

Division 43 Process Gas and Liquid Handling,
Purification, and Storage Equipment

Division 44 Pollution Control Equipment

Division 45 Industry-Specific Manufacturing
Equipment

Division 46 Reserved

Division 47 Reserved

Division 48 Electrical Power Generation

Division 49 Reserved

UNIFORMAT LEVELS AND TITLES

A Substructure

A10   Foundations

A20   Basement Construction

B Shell

B10   Superstructure

B20   Exterior Enclosure

B30   Roofing

C Interiors

C10   Interior Construction

C20   Stairs

C30   Interior Finishes

D Services

D10   Conveying Systems

D20   Plumbing

D30   Heating, Ventilating, and Air Conditioning (HVAC)

D40   Fire Protection Systems

D50   Electrical Systems

E Equipment and Furnishings

E10   Equipment

E20   Furnishings

UNIFORMAT LEVELS AND TITLES *(cont.)*

F Special Construction and Demolition

F10   Special Construction

F20   Selective Demolition

G Building Sitework

G10   Site Preparation

G20   Site Improvements

G30   Site Civil/Mechanical Utilities

G40   Site Electrical Utilities

G90   Other Site Construction

Z General

Z10   General Requirements

Z20   Bidding Requirements, Contract Forms, and Conditions Contingencies

Z90   Project Cost Estimate

Project Description

10   Project description

20   Proposal, bidding and contracting

30   Cost summary

WEIGHT OF CONSTRUCTION MATERIALS

Material	Density (Pounds per Cubic Foot)	Density (Kilograms per Cubic Meter)
Dirt and Mud		
Clay–Dry	100	1602
Clay–Wet	110	1762
Gravel–Dry	105	1682
Gravel–Wet	125	2003
Limestone	160	2563
Loam	80	1282
Loose dirt–Dry	76	1218
Loose dirt–Moist	78	1250
Mud–Flowing	108	1730
Mud–Steady	115	1842
Rock, well blasted	155	2483
Sand–Dry	97	1554
Sand–Wet	119	1906
Metals		
Aluminum	165	2643
Cast iron	450	7209
Gold	1205	19304
Lead	710	11374
Nickel	565	9051
Rolled steel	490	7850
Silver	656	10509
Stainless steel	501	8026
Tin	459	7353
Wrought iron	485	7770
Zinc	440	7049

WEIGHT OF CONSTRUCTION MATERIALS *(cont.)*

Material	Density (Pounds per Cubic Foot)	Density (Kilograms per Cubic Meter)
Wood and Lumber		
Ash	40	641
Aspen	27	433
Birch	43	689
Black oak	41	657
Black walnut	38	609
Cherry	35	561
Chestnut oak	54	865
Chestnut	41	657
Cypress	32	513
Douglas fir	32	513
Eastern fir	25	401
Elm	45	721
Hard maple	43	689
Hemlock	29	465
Hickory	49	785
Ironwood	63	1009
Live oak	59	945
Mahogany	31	497
Oregon pine	32	513
Ponderosa pine	28	449

WEIGHT OF CONSTRUCTION MATERIALS (cont.)

Material	Density (Pounds per Cubic Foot)	Density (Kilograms per Cubic Meter)
Poplar	30	481
Red cedar	23	368
Red oak	41	657
Red pine	30	481
Redwood	26	417
Teak	43	689
White cedar	22	352
White maple	33	529
White oak	46	737
White pine	26	417
White spruce	27	433
White walnut	26	417
Yellow long leaf pine	44	705
Yellow short leaf pine	38	609
Liquids		
Gasoline	42	673
Ice	56	897
Seawater	64	1025
Snow	8	128
Water (near freezing)	63	1009
Water (about to boil)	60	961

EXCAVATION SLOPES

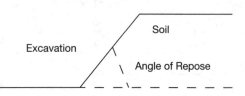

Types of Soil	Maximum Slopes (H:V)	Angle of Repose
Stable Rock	Vertical	90°
Type A (Hard and solid soil)	¾:1	53°
Type B (Soil likely to crack or crumble)	1:1	45°
Type C (Soft, sandy, filled or loose soil)	1½:1	34°

Note
1. Excavation slopes are calculated by dividing horizontal distances by vertical distances. This definition of "excavation slope" is different from as "slopes" in other trades.
2. Data above only apply to excavations less than 20 feet deep.
3. Very few soils are stable enough for "Type A" or better. When no information is available, it is recommended to assume the soil to be Type C.

ESTIMATING BASEMENT EXCAVATION

Estimating Math

Excavation Volume =

(Building Area + Building Perimeter $\times$ Excavation Slope $\times$ Excavation Depth) $\times$ Excavation Depth

Backfill Volume =

Building Perimeter $\times$ Excavation Depth $\times$ Excavation Slope $\times$ Excavation Depth

Estimating Example

A 35' $\times$ 60' building with excavation average depth 9':
Excavation slope is given as 1: 2 (or angle of repose 60°)
for very stable soil

Calculation:

Building Area: 35 $\times$ 60 = 2,100 SF

Building Perimeter: (35 + 60) $\times$ 2 = 190 LF

Excavation Depth: 9 LF

Excavation Volume:

(2100 + 190 $\times$ 0.5 $\times$ 9) $\times$ 9 = 26, 595 CF, or 26, 595/27 = 985 CY

Backfill Volume:

190 $\times$ 0.5 $\times$ 9 $\times$ 9 = 7, 695 CF, or 7, 695/27 = 285 CY

Note: CF = Cubic Foot, CY= Cubic Yard, LF= Linear Foot,
       SF = Square Foot

ESTIMATING TRENCH EXCAVATION

For excavation slope of 1:1 (or angle of repose 45°):

Total cubic yard (CY) of excavation required per linear foot (LF) of trench:

Trench Width (LF)	Trench Depth (LF)								
	2	3	4	5	6	7	8	9	10
1	0.22	0.44	0.74	1.11	1.56	2.07	2.67	3.33	4.07
2	0.30	0.56	0.89	1.30	1.78	2.33	2.96	3.67	4.44
3	0.37	0.67	1.04	1.48	2.00	2.59	3.26	4.00	4.81
4	0.44	0.78	1.19	1.67	2.22	2.85	3.56	4.33	5.19
5	0.52	0.89	1.33	1.85	2.44	3.11	3.85	4.67	5.56
6	0.59	1.00	1.48	2.04	2.67	3.37	4.15	5.00	5.93
7	0.67	1.11	1.63	2.22	2.89	3.63	4.44	5.33	6.30

Total cubic meter of excavation required per meter of trench:

Trench Width (milli-meter)	Trench Depth (millimeter)								
	610	914	1219	1524	1829	2134	2438	2743	3048
305	0.56	1.11	1.86	2.79	3.90	5.20	6.69	8.36	10.22
610	0.74	1.39	2.23	3.25	4.46	5.85	7.43	9.20	11.15
914	0.93	1.67	2.60	3.72	5.02	6.50	8.18	10.03	12.08
1219	1.11	1.95	2.97	4.18	5.57	7.15	8.92	10.87	13.01
1524	1.30	2.23	3.34	4.65	6.13	7.80	9.66	11.71	13.94
1829	1.49	2.51	3.72	5.11	6.69	8.45	10.41	12.54	14.86
2134	1.67	2.79	4.09	5.57	7.25	9.10	11.15	13.38	15.79

ESTIMATING SOIL SWELL

Soil Type	Swell Percentage
Clay	20% to 40%
Earth	20% to 30%
Granite	75% to 80%
Gravel, dry	20% to 30%
Gravel, wet	20% to 30%
Gravel, wet with clay	50% to 60%
Limestone	75% to 80%
Loam	15% to 25%
Quartz	75% to 80%
Rock	40% to 80%
Sand, dry	20% to 30%
Sand, wet	20% to 30%
Sandstone	75% to 80%
Slate	85% to 90%

Estimating Example:

Soil to be hauled away =
In-Place Quantity × (1 + Swell Percentage).

For 1, 000 cubic yards of solid rock excavated based on 50% swell,

You need to haul away: 1000 × (1 + 50%) = 1,500 cubic yards of loose material.

ESTIMATING AGGREGATES

1 ton of aggregate covers:

Thickness	Area
1" deep	240 square feet
2" deep	120 square feet
3" deep	80 square feet
4" deep	60 square feet
5" deep	50 square feet
6" deep	40 square feet
12" deep	20 square feet

1 cubic yard of aggregate covers:

Thickness	Area
1" deep	300 square feet
2" deep	150 square feet
3" deep	100 square feet
4" deep	75 square feet
5" deep	60 square feet
6" deep	50 square feet
12" deep	25 square feet

CONCRETE AND REINFORCING

CONCRETE MIXTURES

Application	Material Volume Ratio		
	Cement	Sand	Gravel
Normal static loads, no rebar; not exposed	1	3.0	6.0
Normal foundations and walls, exposed	1	2.5	5.0
Basement walls	1	2.5	4.0
Waterproof basement walls	1	2.5	3.5
Floors (light duty), driveways, sidewalk	1	2.5	3.0
Steps, driveways, sidewalks	1	2.25	3.0
Reinforced roads, buildings, walls, exposed	1	2.0	4.0
Retaining walls, drive ways	1	2.0	3.5
Swimming pools, fence posts	1	2.0	3.0
Floors (light duty)	1	1.75	4.0
Watertight, reinforced tanks and columns	1	1.5	3.0
High strength columns, girders, floors	1	1.0	2.0
Fence posts	1	1.0	1.5

CONCRETE AND REINFORCING (cont.)

MATERIALS TO MAKE 1 CUBIC YARD OF CONCRETE MIXTURE

Material Volume Ratio			Materials Required		
Cement	Sand	Gravel	Cement (sacks)	Sand (cubic yards)	Gravel (cubic yards)
1	3.0	6.0	4.2	0.47	0.94
1	3.0	5.0	4.6	0.51	0.85
1	2.5	5.0	5.6	0.46	0.92
1	2.5	4.0	5.6	0.52	0.83
1	2.5	3.5	5.9	0.55	0.77
1	2.0	4.0	6.0	0.44	0.89
1	2.0	3.0	7.0	0.52	0.78
1	2.0	2.0	8.2	0.60	0.60
1	1.5	3.0	7.6	0.42	0.85

CONCRETE AND REINFORCING (cont.)

CONCRETE STRENGTH

English (PSI)	Metric (MPa)
2500	18
3000	20
3500	25
4000	30
5000	35
6000	40
7000	43
8000	55

Note: PSI = Pounds per Square Inch,
    MPa = Mega Pascal

REINFORCING BAR WEIGHT

Imperial	Weight (lb/ft)	Former Metric	Current Metric	Weight (kg/m)
#2	0.167	5M	No. 6	0.248
#3	0.376	Not Used	No. 10	0.559
#4	0.668	10M	No. 13	0.994
#5	1.043	15M	No. 16	1.552
#6	1.502	20M	No. 19	2.235
#7	2.044	Not Used	No. 22	3.041
#8	2.670	25M	No. 25	3.973
#9	3.400	30M	No. 29	5.059
#10	4.303	Not Used	No. 32	6.403
#11	5.313	35M	No. 36	7.906
#14	7.650	45M	No. 43	11.383
#18	13.600	55M	No. 57	20.237

WELDED WIRE MESH		
Current mesh name (wire size)	**Former mesh name (wire gauge)**	**Metric name**
2 × 2 W4.0/4.0	2 × 2 – 4/4	50 × 50 MW25.8/25.10
2 × 2 W2.9/2.9	2 × 2 – 6/6	50 × 50 MW18.7/18.9
2 × 2 W2.1/2.1	2 × 2 – 8/8	50 × 50 MW13.3/13.5
2 × 2 W1.4/1.4	2 × 2 – 10/10	50 × 50 MW9.1/9.3
2 × 2 W0.9/0.9	2 × 2 – 12/12	50 × 50 MW5.6/5.8
2 × 2 W0.5/0.5	2 × 2 – 14/14	50 × 50 MW3.2/3.4
2 × 2 W0.3/0.3	2 × 2 – 16/16	50 × 50 MW2.0/2.2
3 × 3 W2.1/2.1	3 × 3 – 8/8	76 × 76 MW13.3/13.5
3 × 3 W1.4/1.4	3 × 3 – 10/10	76 × 76 MW9.1/9.3
3 × 3 W0.9/0.9	3 × 3 – 12/12	76 × 76 MW5.6/5.8
3 × 3 W0.5/0.5	3 × 3 – 14/14	76 × 76 MW3.2/3.4
4 × 4 W4.0/4.0	4 × 4 – 4/4	102 × 102 MW25.8/25.10
4 × 4 W2.9/2.9	4 × 4 – 6/6	102 × 102 MW18.7/18.9
4 × 4 W2.1/2.1	4 × 4 – 8/8	102 × 102 MW13.3/13.5
4 × 4 W1.7/1.7	4 × 4 – 9/9	102 × 102 MW11.1/11.3
4 × 4 W1.4/1.4	4 × 4 – 10/10	102 × 102 MW9.1/9.3
4 × 4 W0.9/0.9	4 × 4 – 12/12	102 × 102 MW5.6/5.8
4 × 4 W0.7/0.7	4 × 4 – 13/13	102 × 102 MW4.2/4.4
4 × 4 W0.5/0.5	4 × 4 – 14/14	102 × 102 MW3.2/3.4
6 × 6 W7.4/7.4	6 × 6 – 0/0	152 × 152 MW47.6/47.8
6 × 6 W6.3/6.3	6 × 6 – 1/1	152 × 152 MW40.6/40.8
6 × 6 W5.4/5.4	6 × 6 – 2/2	152 × 152 MW34.9/34.11
6 × 6 W4.7/4.7	6 × 6 – 3/3	152 × 152 MW30.1/30.3
6 × 6 W4.0/4.0	6 × 6 – 4/4	152 × 152 MW25.8/25.10
6 × 6 W4.0/2.9	6 × 6 – 4/6	152 × 152 MW25.8/18.9
6 × 6 W3.4/3.4	6 × 6 – 5/5	152 × 152 MW21.7/21.9
6 × 6 W2.9/2.9	6 × 6 – 6/6	152 × 152 MW18.7/18.9
6 × 6 W2.5/2.5	6 × 6 – 7/7	152 × 152 MW15.9/15.11
6 × 6 W2.1/2.1	6 × 6 – 8/8	152 × 152 MW13.3/13.5
6 × 6 W1.7/1.7	6 × 6 – 9/9	152 × 152 MW11.1/11.3
6 × 6 W1.4/1.4	6 × 6 – 10/10	152 × 152 MW9.1/9.3
12 × 12 W5.4/5.4	12 × 12 – 2/2	305 × 305 MW34.9/34.11

ESTIMATING CONCRETE CONTINUOUS FOOTING

Estimating Math

Excavation = (Footing Width + 2x Workspace) ×
Excavation Depth × Footing Length

Formwork = (Footing Length + Footing Width) ×
Footing Height × 2

Concrete = Footing Length × Footing Width ×
Footing Height

Continuous Rebar = Footing Length × Count of Bars

Transverse Rebar = (Footing Length/Bar Spacing +1) ×
Footing Width × Count of Bars

Dowels = (Footing Length/Dowel Spacing +1) ×
Dowel Length × Count of Dowels

Backfill = Excavation Volume − Concrete Volume

Note:

1. Convert units as appropriate, e.g. 1 foot = 12 inches,
   1 cubic yard = 27 cubic feet.
2. The formula above does not allow excavation
   volumes for different slopes. Please make
   adjustments as necessary.
3. Rebar lengths are to be multiplied by pounds per linear
   foot (decided by bar number) to obtain the weight.
4. Quantities are net and approximate. Verify with site
   conditions and add 5% to 15% waste as necessary.
5. Refer to quantity-takeoff check lists in previous
   chapters for more information.

ESTIMATING CONCRETE CONTINUOUS FOOTING *(cont.)*

Cubic Yard (or Cubic Meter) of Concrete per Linear Foot (or Meter) of Footer

Footing Width (in.)	Footing Depth (in.)	Concrete Volume Per LF of Footing (CY)	Footing Width (mm.)	Footing Depth (mm.)	Concrete Volume Per Meter of Footing (M³)
12"	6"	0.019	305	152	0.046
12"	8"	0.025	305	203	0.062
12"	10"	0.031	305	254	0.077
12"	12"	0.037	305	305	0.093
12"	18"	0.056	305	457	0.139
16"	8"	0.033	406	203	0.083
16"	10"	0.041	406	254	0.103
16"	12"	0.049	406	305	0.124
18"	8"	0.037	457	203	0.093
18"	10"	0.046	457	254	0.116
18"	24"	0.111	457	610	0.279
20"	12"	0.062	508	305	0.155
24"	12"	0.074	610	305	0.186
24"	24"	0.148	610	610	0.372
24"	30"	0.185	610	762	0.465
24"	36"	0.222	610	914	0.557
30"	36"	0.278	762	914	0.697
30"	42"	0.324	762	1067	0.813
36"	48"	0.444	914	1219	1.115

Note: Quantities are net. Add waste 5% to 15%

Estimating Example:

1. For 200' footing 12" wide, 18" high, the required concrete is
   0.056 CY/LF × 200 LF = 11.2 CY
   Add 15% waste, the result is: 11.2 × (1+15%) = 12.9 CY

2. For 100m footing 406mm wide, 305mm high, the required
   concrete is 0.124m³/m × 100m = 12.4m³
   Add 15% waste, the result is: 12.4 x (1+15%) = 14.3m³

ESTIMATING CONCRETE SPREAD FOOTINGS

Estimating Math

Excavation= (Footing Length + 2x Workspace) × (Footing Width + 2x Workspace) × Footing Height

Formwork = Count of Footings × (Footing Length + Footing Width) × 2 × Footing Height

Concrete = Count of Footings × Footing Length × Footing Width × Footing Height

Rebar = Count of Footings × (Footing Length + Footing Width) × Count of Bars

Dowel = Count of Footings × Dowel Length × Count of Dowels

Dowel Ties = Count of Footings × (Dowel Length/Tie Spacing +1) × (Footing Length + Footing Width) × 2

Backfill = Excavation Volume − Concrete Volume

Note:

1. Convert units as appropriate, e.g., 1 foot = 12 inches, 1 cubic yard = 27 cubic feet.
2. The formula above does not allow excavation volumes for different slopes. Please make adjustments as necessary.
3. Rebar lengths are to be multiplied by pounds per linear foot (decided by bar number) to obtain the weight.
4. Quantities are net and approximate. Verify with site conditions and add 5% to 15% waste as necessary.
5. Refer to quantity-takeoff check lists in previous chapters for more information.

ESTIMATING CONCRETE WALLS

Estimating Math

Formwork = Wall Length × Wall Height × 2 +
            Wall Thickness × Wall Height × 2

Concrete = Wall Length × Wall Height × Wall Thickness

Horizontal Rebar = Wall Length × Count of Bars

Vertical Rebar = (Wall Length/Bar Spacing +1) ×
            Wall Height × Count of Bars

Dowels = (Wall Length /Dowel Spacing +1) ×
            Dowel Length × Count of Dowels

Insulation = Wall Length × Wall Height

Dampproofing = Wall Length × Wall Height

Note:

1. Convert units as appropriate, e.g., 1 foot = 12 inches,
   1 cubic yard = 27 cubic feet.
2. See footing estimating to see if you need additional
   excavation and backfill.
3. Rebar lengths are to be multiplied by pounds per linear
   foot (determined by bar number) to obtain the weight.
4. Quantities are net and approximate. Verify with site
   conditions and add 5% to 15% waste as necessary.
5. Refer to quantity-takeoff check lists in previous
   chapters for more information.

ESTIMATING CONCRETE WALLS *(cont.)*

Cubic Yard (or Cubic Meter) of Concrete per Square Foot (or Square Meter) of Wall

Wall Thickness (in)	Concrete Volume Per SF of Wall (CY)	Wall Thickness (mm)	Concrete Volume Per M^2 of Wall (M^3)
4"	0.012	102	0.102
6"	0.019	152	0.152
8"	0.025	203	0.203
10"	0.031	254	0.254
12"	0.037	305	0.305

Note: Quantities are net. Add 5% to 15% waste.
Estimating Example:

1. For 500 SF of 8" thick foundation wall, the required concrete is:
   0.025 CY/SF × 500 SF = 12.5 CY
   Add 15% waste, the result is: 12.5 × (1 + 15%) = 14.4 CY

2. For 100m^2 of 305mm thick foundation wall, the required concrete is:
   0.305m^3/m^2 × 100m^2 = 30.5m^3
   Add 15% waste, the result is: 30.5 × (1 + 15%) = 35.1m^3

ESTIMATING CONCRETE SLABS

Estimating Math

Gravel Base = Slab Width × Slab Length

Formwork = (Slab Width + Slab Length) × 2 × Slab Thickness

Concrete = Slab Width × Slab Length × Slab Thickness

Rebar = (Slab Width/Rebar Spacing + 1) × Slab Length + (Slab Length/Rebar Spacing + 1) × Slab Width

Wire Mesh = Slab Width × Slab Length /Roll Coverage

Vapor Barrier = Slab Width × Slab Length

Control Joints = (Slab Length /Joint Spacing + 1) × Joint Length

Note

1. Convert units as appropriate, e.g., 1 foot = 12 inches, 1 cubic yard = 27 cubic feet.
2. The above formulas apply to concrete slabs in rectangular shape and reinforced in both directions.
3. Rebar lengths will be multiplied by pounds per linear foot (determined by bar number) to obtain the weight.
4. Quantities are net and approximate. Verify with site conditions and add 5% to 15% waste as necessary.
5. Refer to quantity-takeoff check lists in previous chapters for more information.

ESTIMATING CONCRETE SLABS *(cont.)*

Cubic Yard (or Cubic Meter) of Concrete per Square Foot (or Square Meter) of Slab

Slab Thickness (in.)	Concrete Volume Per SF of Slab (CY)	Slab Thickness (mm.)	Concrete Volume Per M² of Wall (M³)
1"	0.003	25	0.025
1½"	0.005	38	0.038
2"	0.006	51	0.051
2½"	0.008	64	0.064
3"	0.009	76	0.076
3½"	0.011	89	0.089
4"	0.012	102	0.102
4½"	0.014	114	0.114
5"	0.015	127	0.127
5½"	0.017	140	0.140
6"	0.019	152	0.152
6½"	0.020	165	0.165
7"	0.022	178	0.178
7½"	0.023	191	0.191
8"	0.025	203	0.203
8½"	0.026	216	0.216
9"	0.028	229	0.229
9½"	0.029	241	0.241
10"	0.031	254	0.254
10½"	0.032	267	0.267
11"	0.003	279	0.279
11½"	0.005	292	0.292
12"	0.037	305	0.305

Note: Quantities are net. Add 5% to 15% waste.
Estimating Example:
1. For 500 SF of 8" thick slab, the required concrete is
   0.025 CY/SF × 500 SF = 12.5 CY
   Add 15% waste, the result is: 12.5 × (1 + 15%) = 14.4 CY
2. For 200m² of 152mm thick slab, the required concrete is:
   0.152m³/m² × 200m² = 30.4m³
   Add 15% waste, the result is: 30.4 × (1 + 15%) = 35m³

ESTIMATING CONCRETE COLUMNS

Estimating Math

Formwork = Column Section Perimeter × Column Height

Concrete = Column Section Area × Column Height

Vertical Rebar = Column Height × Count of Bars

Rebar Ties = (Column Height/ Tie Spacing + 1) ×
Tie Length per Section

For Round Columns:

Formwork = 3.1416 × Column Section Diameter ×
Column Height

Concrete = 0.7854 × Column Section Diameter ×
Column Section Diameter × Column Height

For Square Columns:

Formwork = (Column Section Length + Column Section
Width) × 2 × Column Height

Concrete = Column Section Length × Column Section
Width × Column Height

Note:

1. Convert units as appropriate, e.g., 1 foot = 12 inches,
   1 cubic yard = 27 cubic feet.
2. Rebar lengths will be multiplied by pounds per linear
   foot (determined by bar number) to obtain the weight.
3. Quantities are net and approximate. Verify with site
   conditions and add 5% to 15% waste as necessary.
4. Refer to quantity-takeoff check lists in previous
   chapters for more information.

ESTIMATING CONCRETE COLUMNS *(cont.)*

Cubic Yard (or Cubic Meter) of Concrete per Linear Foot (or Meter) of Round Column

Column Diameter (in.)	Concrete Volume Per LF of Column (CY)	Column Diameter (mm.)	Concrete Volume Per Meter of Column (M³)
12"	0.029	305	0.073
13"	0.034	330	0.086
14"	0.040	356	0.099
15"	0.045	381	0.114
16"	0.052	406	0.130
17"	0.058	432	0.146
18"	0.065	457	0.164
19"	0.073	483	0.183
20"	0.081	508	0.203
21"	0.089	533	0.223
22"	0.098	559	0.245
23"	0.107	584	0.268
24"	0.116	610	0.292
25"	0.126	635	0.317
26"	0.137	660	0.343
27"	0.147	686	0.369
28"	0.158	711	0.397
29"	0.170	737	0.426
30"	0.182	762	0.456
31"	0.194	787	0.487
32"	0.207	813	0.519
33"	0.220	838	0.552
34"	0.234	864	0.586
35"	0.247	889	0.621
36"	0.262	914	0.657

Note: Quantities are net. Add 5% to 15% waste.

Estimating Example:
1. For one 12" round column 20' high, the required concrete is
   0.029 CY/LF × 20 LF = 0.58 CY
   Add 15% waste, the result is: 0.58 × (1 + 15%) = 0.67 CY
2. For one 610mm round column 10m high, the required concrete is:
   0.292m³/m × 10m = 2.92m³
   Add 15% waste, the result is: 2.92 × (1 + 15%) = 3.36m³

ESTIMATING CONCRETE COLUMNS (cont.)

Cubic Yard (or Cubic Meter) of Concrete per Linear Foot (or Meter) of Square Column

Column Size (in. × in.)	Concrete Volume Per LF of Column (CY)	Column Size (mm. × mm.)	Concrete Volume Per Meter of Column (M³)
12" × 12"	0.037	305 × 305	0.093
13" × 13"	0.043	330 × 330	0.109
14" × 14"	0.050	356 × 356	0.126
15" × 15"	0.058	381 × 381	0.145
16" × 16"	0.066	406 × 406	0.165
17" × 17"	0.074	432 × 432	0.186
18" × 18"	0.083	457 × 457	0.209
19" × 19"	0.093	483 × 483	0.233
20" × 20"	0.103	508 × 508	0.258
21" × 21"	0.113	533 × 533	0.285
22" × 22"	0.124	559 × 559	0.312
23" × 23"	0.136	584 × 584	0.341
24" × 24"	0.148	610 × 610	0.372
25" × 25"	0.161	635 × 635	0.403
26" × 26"	0.174	660 × 660	0.436
27" × 27"	0.188	686 × 686	0.470
28" × 28"	0.202	712 × 712	0.506
29" × 29"	0.216	737 × 737	0.543
30" × 30"	0.231	762 × 762	0.581
31" × 31"	0.247	787 × 787	0.620
32" × 32"	0.263	813 × 813	0.661
33" × 33"	0.280	839 × 839	0.703
34" × 34"	0.297	864 × 864	0.746
35" × 35"	0.315	889 × 889	0.790
36" × 36"	0.333	914 × 914	0.836

Note: Quantities are net. Add 5% to 15% waste.

Estimating Example:

1. For one 20" × 20" square column 30' high, the required concrete is
   0.103 CY/LF × 30 LF = 3.09 CY

   Add 15% waste, the result is: 3.09 × (1 + 15%) = 3.55 CY

2. For one 508 × 508mm square column 5m high, the required concrete is:
   0.258m³/m × 5m = 1.29m³

   Add 15% waste, the result is: 1.29 × (1 + 15%) = 1.48m³

ESTIMATING CONCRETE BEAMS

Estimating Math

Formwork = Beam Height $\times$ Beam Length $\times$ 2 +
      Beam Width $\times$ Beam Length

Concrete = Beam Height $\times$ Beam Width $\times$ Beam Length

Continuous Rebar = Beam Length $\times$ Count of Bars

Rebar Ties = (Beam Length / Tie Spacing + 1) $\times$
      Tie Length per Section

Note

1. Convert units as appropriate, e.g., 1 foot = 12 inches,
   1 cubic yard = 27 cubic feet.

2. The above formula includes formwork for rectangular
   beams on sides and bottom.

3. Rebar lengths will be multiplied by pounds per linear
   foot (determined by bar number) to obtain the weight.

4. Quantities are net and approximate. Verify with site
   conditions and add 5% to 15% waste as necessary.

5. Refer to quantity-takeoff check lists in previous
   chapters for more information.

ESTIMATING CONCRETE BEAMS (*cont.*)

Cubic Yard (or Cubic Meter) of Concrete per Linear Foot (or Meter) of Beam

Beam Size (in. × in.)	Concrete Volume Per LF of Beam (CY)	Beam Size (mm. × mm.)	Concrete Volume Per Meter of Beam (M³)
8" × 8"	0.016	203 × 203	0.041
8" × 12"	0.025	203 × 305	0.062
8" × 16"	0.033	203 × 406	0.083
8" × 20"	0.041	203 × 508	0.103
8" × 24"	0.049	203 × 610	0.124
12" × 12"	0.037	305 × 305	0.093
12" × 16"	0.049	305 × 406	0.124
12" × 20"	0.062	305 × 508	0.155
12" × 24"	0.074	305 × 610	0.186

Note: Quantities are net. Add 5% to 15% waste.

Estimating Example:

1. For 8" × 16" beam 100' long, the required concrete is
   0.033 CY/LF × 100 LF = 3.3 CY
   Add 15% waste, the result is: 3.3 × (1 + 15%) = 3.8 CY

2. For 305 × 610mm beam 150m long, the required concrete is:
   0.186m³/m × 150m = 27.9m³
   Add 15% waste, the result is: 27.9 × (1 + 15%) = 32.1m³

MASONRY BLOCKS, BRICKS AND MORTAR

Typical CMU (Concrete Masonry Unit) Dimensions

Imperial		Metric	
Nominal Size (inch)	Actual Size (inch)	Nominal Size (millimeter)	Actual Size (millimeter)
Standard Block			
$4 \times 8 \times 16$	$3\frac{5}{8} \times 7\frac{5}{8} \times 15\frac{5}{8}$	$102 \times 203 \times 406$	$92 \times 194 \times 397$
$6 \times 8 \times 16$	$5\frac{5}{8} \times 7\frac{5}{8} \times 15\frac{5}{8}$	$152 \times 203 \times 406$	$143 \times 194 \times 397$
$8 \times 8 \times 16$	$7\frac{5}{8} \times 7\frac{5}{8} \times 15\frac{5}{8}$	$203 \times 203 \times 406$	$194 \times 194 \times 397$
$10 \times 8 \times 16$	$9\frac{5}{8} \times 7\frac{5}{8} \times 15\frac{5}{8}$	$254 \times 203 \times 406$	$244 \times 194 \times 397$
$12 \times 8 \times 16$	$11\frac{5}{8} \times 7\frac{5}{8} \times 15\frac{5}{8}$	$305 \times 203 \times 406$	$295 \times 194 \times 397$
Half Block			
$4 \times 8 \times 8$	$3\frac{5}{8} \times 7\frac{5}{8} \times 7\frac{5}{8}$	$102 \times 203 \times 203$	$92 \times 194 \times 194$
$6 \times 8 \times 8$	$5\frac{5}{8} \times 7\frac{5}{8} \times 7\frac{5}{8}$	$152 \times 203 \times 203$	$143 \times 194 \times 194$
$8 \times 8 \times 8$	$7\frac{5}{8} \times 7\frac{5}{8} \times 7\frac{5}{8}$	$203 \times 203 \times 203$	$194 \times 194 \times 194$
$12 \times 8 \times 8$	$11\frac{5}{8} \times 7\frac{5}{8} \times 7\frac{5}{8}$	$305 \times 203 \times 203$	$295 \times 194 \times 194$

MASONRY BLOCKS, BRICKS AND MORTAR (cont.)

Face Brick Sizes

Type	Nominal Size (inch)	Actual Size (inch)	Quantity per Square Foot
Standard	$4 \times 2\frac{2}{3} \times 8$	$3\frac{5}{8} \times 2\frac{1}{4} \times 8$	6.27
Modular	$4 \times 2\frac{2}{3} \times 8$	$3\frac{5}{8} \times 2\frac{1}{4} \times 7\frac{5}{8}$	6.86
King	$3\frac{3}{8} \times 3 \times 10$	$3 \times 2\frac{5}{8} \times 9\frac{5}{8}$	4.8
Queen	$2\frac{3}{4} \times 3 \times 10$	$3\frac{1}{8} \times 2\frac{3}{4} \times 9\frac{5}{8}$	4.61
Engineer	$4 \times 3\frac{1}{5} \times 8$	$3\frac{5}{8} \times 2\frac{13}{16} \times 7\frac{5}{8}$	5.65
Economy	$4 \times 4 \times 8$	$3\frac{5}{8} \times 3\frac{5}{8} \times 7\frac{5}{8}$	4.5
Utility	$4 \times 4 \times 12$	$3\frac{5}{8} \times 3\frac{5}{8} \times 11\frac{1}{2}$	3.03
Jumbo	$4 \times 3 \times 8$	$3\frac{5}{8} \times 2\frac{3}{4} \times 8$	5.5
Norman	$4 \times 2\frac{2}{3} \times 12$	$3\frac{5}{8} \times 2\frac{1}{4} \times 11\frac{5}{8}$	4.57
Norwegian	$3\frac{1}{2} \times 3 \times 12$	$3\frac{1}{2} \times 2\frac{3}{4} \times 11\frac{5}{8}$	3.84

MASONRY BLOCKS, BRICKS AND MORTAR *(cont.)*

Bond Adjustment Factors for Face Bricks

Bond Type	Percentage
English bond (with a full header course every other course)	50.00%
English bond (with a full header course every 6th course)	16.67%
English cross bond (with a full header course every other course)	50.00%
English cross bond (with a full header course every 6th course)	16.67%
Common bond (with a full header course every 5th course)	20.00%
Common bond (with a full header course every 6th course)	16.67%
Common bond (with a full header course every 7th course)	14.33%
Dutch bond (with a full header course every other course)	50.00%
Dutch bond (with a full header course every 6th course)	16.67%
Dutch cross bond (with a full header course every other course)	50.00%
Dutch cross bond (with a full header course every 6th course)	16.67%
Double header bond (two headers and a stretcher every 5th course)	10.00%
Double header bond (two headers and a stretcher every 6th course)	8.33%
Flemish bond (with a full header course every other course)	33.33%
Flemish bond (with a full header course every 6th course)	5.60%
Double Flemish bond (with a full header course every other course)	10.00%
Double Flemish bond (with a full header course every 3rd course)	6.67%
3 stretcher Flemish bond (with a full header course every other course)	7.14%
3 stretcher Flemish bond (with a full header course every 3rd course)	4.80%
4 stretcher Flemish bond (with a full header course every other course)	5.60%
4 stretcher Flemish bond (with a full header course every 3rd course)	3.70%

Note:
Percentages refer to what needs to be added to quantities normally calculated for running bond walls.

Example:
The bricks calculated based on running bond is 2000 EA.
The wall is actually laid in common bond with a full header course every 5th course.
The adjusted brick quantity is 2000 x (1 + 20%) = 2400 EA.

MASONRY BLOCKS, BRICKS AND MORTAR (cont.)

Mortar Types and Applications

Type	Applications	28 Day Strength
M	Foundation masonry, retaining walls, sidewalks, sewers, manholes or situations in contact with ground or below grade	4900 to 5400 PSI
S	Reinforced masonry, cavity walls exposed to strong winds, ceramic veneers	2100 to 2800 PSI
N	Exposed masonry above grade	800 to 1200 PSI
O	Interior low strength load bearing wall	less than 100 PSI

Mortar Material Volume Ratio

	Portland Mix			Masonry Mix		
Type	Portland Cement	Hydrated Lime	Sand	Portland Cement	Masonry Cement	Sand
M	1	1/4	3	1	1	6
S	1	1/2	4.5	1/2	1	4.5
N	1	1	6	N/A	1	3
O	1	2	9	N/A	1	3

ESTIMATING MASONRY

Masonry Quantity Take-Off Procedures

1. Identify material requirements, such as block/brick type and manufacturer; rebar lap; cell-fill concrete strength; block fire resistance requirements, etc.

2. From foundation plan, take off stem wall along the building perimeter.

3. Compare roof framing plan with exterior elevations. Take off the main exterior wall above grade.

4. Find masonry beams on roofs, parapets and over openings.

5. Add half blocks around openings, concrete tie columns, and wall control joints.

6. Figure corner blocks as necessary.

7. Check building interior for load-bearing masonry wall.

8. Do not forget building exterior veneer and columns (e.g., split face blocks, face bricks, and cast stone, etc.)

9. From site plan, find dumpsters and screen walls.

10. Take off structural pre-cast lintels per their dimensions.

11. Take off architectural pre-cast like column caps, water table, and window sills.

MASONRY ESTIMATING RULES OF THUMB

Item	Quantities
Blocks	
Standard Block	1.125 block per SF of wall area
Half Block	2.25 block per SF of wall area
Bricks	
Face Brick Modular	7.0 brick per SF of wall area
Oversize Brick	6.0 brick per SF of wall area
Utility Brick	3.0 brick per SF of wall area
Mortar	
Block	3 bags per 100 block
Face Brick Modular	7 bags per 1000 brick
Oversized Brick	8 bags per 1000 brick
Utility Brick	10 bags per 1000 brick
Sand	
Sand	1 CY per 7 bags mortar
Cell-fill Concrete Quantities	
$6 \times 8 \times 16$	0.17 CF/block
$8 \times 8 \times 16$	0.25 CF/block
$10 \times 8 \times 16$	0.33 CF/block
$12 \times 8 \times 16$	0.39 CF/block
$6 \times 8 \times 16$ Bond Beam	0.173 CF per LF
$8 \times 8 \times 16$ Bond Beam	0.22 CF per LF
$8 \times 8 \times 16$ Deep Bond Beam	0.46 CF per LF
$12 \times 8 \times 16$ Bond Beam	0.37 CF per LF
$12 \times 8 \times 16$ Deep Bond Beam	0.74 CF per LF

ESTIMATING STRUCTURAL STEEL

Estimating Structural Shapes and Joists
1. Divide the building into smaller portions of floor areas or bays
2. Identify different shapes and section sizes
3. Measure and sum up lengths for each shape separately
4. Multiply lengths by the unit weight per linear foot for each shape
5. Sum up the total weight, figure connection and waste (10 to 20%)
6. Convert the result to tons

Estimating Metal Decking
1. Separate different types and gauges of decking
2. Measure flat area by SF (square feet)
3. Add overhang area
4. Allow for pitch
5. Allow for side and end overlap
6. Convert the total area to SQ (100 SF)

Example:
Calculate the total weight of the following shapes:

Designation	Length	Count
W 12 × 14	15' 6⅝"	4
S 5 × 10	20' 7¾"	3
HSS 20 × 20 × ⅜	25' 2⅜"	2

Linear feet of W 12 × 14 = 4 × 15' 6⅝" = 4 × 15.55' = 62.20'
Linear feet of S 5 × 10 = 3 × 20' 7¾"= 3 × 20.64' = 61.92'
Linear feet of HSS 20 × 20 × ⅜ = 2 × 25' 2⅜"= 2 × 25.20' = 50.40'
Look up their unit weights (Pounds per linear foot) from the weight tables in this book:
Pounds of W 12 × 14 = 62.20' × 14 lbs/ft = 871 lbs
Pounds of S 5 × 10 = 61.92' × 10 lbs/ft = 620 lbs
Pounds of HSS 20 × 20 × ⅜ = 50.40' × 103.22 lbs/ft = 5,203 lbs

Total Pounds: 871 + 620 + 5,203 = 6,693 lbs
Convert to Tons (short ton): 6,693/2000 = 3.35 Tons

STRUCTURAL STEEL WEIGHT TABLE
(IMPERIAL AND METRIC)

STEEL JOISTS (K SERIES)

Designation	Imperial		Metric	
	Depth (in)	Weight (lb/ft)	Depth (mm)	Weight (kg/m)
8K1	8	5.1	203	7.6
10K1	10	5.0	254	7.4
12K1	12	5.0	305	7.4
12K3	12	5.7	305	8.5
12K5	12	7.1	305	10.6
14K1	14	5.2	356	7.7
14K3	14	6.0	356	8.9
14K4	14	6.7	356	10.0
14K6	14	7.7	356	11.5
16K2	16	5.5	406	8.2
16K3	16	6.3	406	9.4
16K4	16	7.0	406	10.4
16K5	16	7.5	406	11.2
16K6	16	8.1	406	12.1
16K7	16	8.6	406	12.8
16K9	16	10.0	406	14.9
18K3	18	6.6	457	9.8
18K4	18	7.2	457	10.7
18K5	18	7.7	457	11.5
18K6	18	8.5	457	12.6
18K7	18	9	457	13.4

STRUCTURAL STEEL WEIGHT TABLE (IMPERIAL AND METRIC) (cont.)

STEEL JOISTS (K SERIES)

Designation	Imperial		Metric	
	Depth (in)	Weight (lb/ft)	Depth (mm)	Weight (kg/m)
18K9	18	10.2	457	15.2
18K10	18	11.7	457	17.4
20K3	20	6.7	508	10.0
20K4	20	7.6	508	11.3
20K5	20	8.2	508	12.2
20K6	20	8.9	508	13.2
20K7	20	9.3	508	13.8
20K9	20	10.8	508	16.1
20K10	20	12.2	508	18.2
22K4	22	8	559	11.9
22K5	22	8.8	559	13.1
22K6	22	9.2	559	13.7
22K7	22	9.7	559	14.4
22K9	22	11.3	559	16.8
22K10	22	12.6	559	18.8
22K11	22	13.8	559	20.5
24K4	24	8.4	610	12.5
24K5	24	9.3	610	13.8
24K6	24	9.7	610	14.4
24K7	24	10.1	610	15.0
24K8	24	11.5	610	17.1

STRUCTURAL STEEL WEIGHT TABLE
(IMPERIAL AND METRIC) *(cont.)*

STEEL JOISTS (K SERIES)

Designation	Imperial		Metric	
	Depth (in)	Weight (lb/ft)	Depth (mm)	Weight (kg/m)
24K9	24	12	610	17.9
24K10	24	13.1	610	19.5
24K12	24	16	610	23.8
26K5	26	9.8	660	14.6
26K6	26	10.6	660	15.8
26K7	26	10.9	660	16.2
26K8	26	12.1	660	18.0
26K9	26	12.2	660	18.2
26K10	26	13.8	660	20.5
26K12	26	16.6	660	24.7
28K6	28	11.4	711	17.0
28K7	28	11.8	711	17.6
28K8	28	12.7	711	18.9
28K9	28	13	711	19.3
28K10	28	14.3	711	21.3
28K12	28	17.1	711	25.5
30K7	30	12.3	762	18.3
30K8	30	13.2	762	19.6
30K9	30	13.4	762	19.9
30K10	30	15	762	22.3
30K11	30	16.4	762	24.4
30K12	30	17.6	762	26.2

STRUCTURAL STEEL WEIGHT TABLE
(IMPERIAL AND METRIC) *(cont.)*

STEEL JOISTS (CS SERIES)

Designation	Imperial		Metric	
	Depth (in)	Weight (lb/ft)	Depth (mm)	Weight (kg/m)
10CS1	10	6.0	254	8.9
10CS2	10	7.5	254	11.2
10CS3	10	10.0	254	14.9
12CS1	12	6.0	305	8.9
12CS2	12	8.0	305	11.9
12CS3	12	10.0	305	14.9
14CS1	14	6.5	356	9.7
14CS2	14	8.0	356	11.9
14CS3	14	10.0	356	14.9
16CS2	16	8.5	406	12.7
16CS3	16	10.5	406	15.6
16CS4	16	14.5	406	21.6
16CS5	16	18.0	406	26.8
18CS2	18	9.0	457	13.4
18CS3	18	11.0	457	16.4
18CS4	18	15.0	457	22.4
18CS5	18	18.5	457	27.6
20CS2	20	9.5	508	14.2
20CS3	20	11.5	508	17.1
20CS4	20	16.5	508	24.6
20CS5	20	20.0	508	29.8

STRUCTURAL STEEL WEIGHT TABLE
(IMPERIAL AND METRIC) *(cont.)*

STEEL JOISTS (CS SERIES)

Designation	Imperial		Metric	
	Depth (in)	Weight (lb/ft)	Depth (mm)	Weight (kg/m)
22CS2	22	10.0	559	14.9
22CS3	22	12.5	559	18.6
22CS4	22	16.5	559	24.6
22CS5	22	20.5	559	30.5
24CS2	24	10.0	610	14.9
24CS3	24	12.5	610	18.6
24CS4	24	16.5	610	24.6
24CS5	24	20.5	610	30.5
26CS2	26	10.0	660	14.9
26CS3	26	12.5	660	18.6
26CS4	26	16.5	660	24.6
26CS5	26	20.5	660	30.5
28CS2	28	10.5	711	15.6
28CS3	28	12.5	711	18.6
28CS4	28	16.5	711	24.6
28CS5	28	20.5	711	30.5
30CS2	30	11.0	762	16.4
30CS3	30	13.0	762	19.4
30CS4	30	16.5	762	24.6
30CS5	30	21.0	762	31.3

STRUCTURAL STEEL WEIGHT TABLE
(IMPERIAL AND METRIC) *(cont.)*

STEEL JOISTS (LH SERIES)

Designation	Imperial		Metric	
	Depth (in)	Weight (lb/ft)	Depth (mm)	Weight (kg/m)
18LH02	18	10	457	15
18LH03	18	11	457	16
18LH04	18	12	457	18
18LH05	18	15	457	22
18LH06	18	15	457	22
18LH07	18	17	457	25
18LH08	18	19	457	28
18LH09	18	21	457	31
20LH02	20	10	508	15
20LH03	20	11	508	16
20LH04	20	12	508	18
20LH05	20	14	508	21
20LH06	20	15	508	22
20LH07	20	17	508	25
20LH08	20	19	508	28
20LH09	20	21	508	31
20LH10	20	23	508	34
24LH03	24	11	610	16
24LH04	24	12	610	18
24LH05	24	13	610	19
24LH06	24	16	610	24
24LH07	24	17	610	25
24LH08	24	18	610	27
24LH09	24	21	610	31
24LH10	24	23	610	34
24LH11	24	25	610	37
28LH05	28	13	711	19

STRUCTURAL STEEL WEIGHT TABLE
(IMPERIAL AND METRIC) *(cont.)*

STEEL JOISTS (LH SERIES)

Designation	Imperial		Metric	
	Depth (in)	Weight (lb/ft)	Depth (mm)	Weight (kg/m)
28LH06	28	16	711	24
28LH07	28	17	711	25
28LH08	28	18	711	27
28LH09	28	21	711	31
28LH10	28	23	711	34
28LH11	28	25	711	37
28LH12	28	27	711	40
28LH13	28	30	711	45
32LH06	32	14	813	21
32LH07	32	16	813	24
32LH08	32	17	813	25
32LH09	32	21	813	31
32LH10	32	21	813	31
32LH11	32	24	813	36
32LH12	32	27	813	40
32LH13	32	30	813	45
32LH14	32	33	813	49
32LH15	32	35	813	52
36LH07	36	16	914	24
36LH08	36	18	914	27
36LH09	36	21	914	31
36LH10	36	21	914	31
36LH11	36	23	914	34
36LH12	36	25	914	37
36LH13	36	30	914	45
36LH14	36	36	914	54
36LH15	36	36	914	54

STRUCTURAL STEEL WEIGHT TABLE
(IMPERIAL AND METRIC) *(cont.)*

STEEL JOISTS (LH SERIES)

Designation	Imperial		Metric	
	Depth (in)	Weight (lb/ft)	Depth (mm)	Weight (kg/m)
40LH08	40	16	1016	24
40LH09	40	21	1016	31
40LH10	40	21	1016	31
40LH11	40	22	1016	33
40LH12	40	25	1016	37
40LH13	40	30	1016	45
40LH14	40	35	1016	52
40LH15	40	36	1016	54
40LH16	40	42	1016	63
44LH09	44	19	1118	28
44LH10	44	21	1118	31
44LH11	44	22	1118	33
44LH12	44	25	1118	37
44LH13	44	30	1118	45
44LH14	44	31	1118	46
44LH15	44	36	1118	54
44LH16	44	42	1118	63
44LH17	44	47	1118	70
48LH10	48	21	1219	31
48LH11	48	22	1219	33
48LH12	48	25	1219	37
48LH13	48	29	1219	43
48LH14	48	32	1219	48
48LH15	48	36	1219	54
48LH16	48	42	1219	63
48LH17	48	47	1219	70

STRUCTURAL STEEL WEIGHT TABLE
(IMPERIAL AND METRIC) *(cont.)*

STEEL JOISTS (DLH SERIES)

Designation	Imperial		Metric	
	Depth (in)	Weight (lb/ft)	Depth (mm)	Weight (kg/m)
52DLH10	52	25	1321	37
52DLH11	52	26	1321	39
52DLH12	52	29	1321	43
52DLH13	52	34	1321	51
52DLH14	52	39	1321	58
52DLH15	52	42	1321	63
52DLH16	52	45	1321	67
52DLH17	52	52	1321	77
56DLH11	56	26	1422	39
56DLH12	56	30	1422	45
56DLH13	56	34	1422	51
56DLH14	56	39	1422	58
56DLH15	56	42	1422	63
56DLH16	56	46	1422	68
56DLH17	56	51	1422	76
60DLH12	60	29	1524	43
60DLH13	60	35	1524	52
60DLH14	60	40	1524	60
60DLH15	60	43	1524	64
60DLH16	60	46	1524	68
60DLH17	60	52	1524	77

STRUCTURAL STEEL WEIGHT TABLE
(IMPERIAL AND METRIC) *(cont.)*

STEEL JOISTS (DLH SERIES)

Designation	Imperial		Metric	
	Depth (in)	Weight (lb/ft)	Depth (mm)	Weight (kg/m)
60DLH18	60	59	1524	88
64DLH12	64	31	1626	46
64DLH13	64	34	1626	51
64DLH14	64	40	1626	60
64DLH15	64	43	1626	64
64DLH16	64	46	1626	68
64DLH17	64	52	1626	77
64DLH18	64	59	1626	88
69DLH13	68	37	1727	55
69DLH14	68	40	1727	60
69DLH15	68	40	1727	65
69DLH16	68	49	1727	73
69DLH17	68	55	1727	82
69DLH18	68	61	1727	91
69DLH19	68	67	1727	100
72DLH14	72	41	1829	61
72DLH15	72	44	1829	65
72DLH16	72	50	1829	74
72DLH17	72	56	1829	83
72DLH18	72	59	1829	88
72DLH19	72	70	1829	104

STRUCTURAL STEEL WEIGHT TABLE
(IMPERIAL AND METRIC) *(cont.)*

STEEL JOISTS (SLH SERIES)

Designation	Imperial		Metric	
	Depth (in)	Weight (lb/ft)	Depth (mm)	Weight (kg/m)
80SLH15	80	40	2032	60
80SLH16	80	46	2032	69
80SLH17	80	53	2032	79
80SLH18	80	60	2032	89
80SLH19	80	67	2032	100
80SLH20	80	75	2032	112
88SLH16	88	46	2235	69
88SLH17	88	51	2235	76
88SLH18	88	58	2235	86
88SLH19	88	65	2235	97
88SLH20	88	76	2235	113
88SLH21	88	89	2235	133
96SLH17	96	52	2438	77
96SLH18	96	58	2438	86
96SLH19	96	66	2438	98
96SLH20	96	74	2438	110
96SLH21	96	90	2438	134
96SLH22	96	102	2438	152

STRUCTURAL STEEL WEIGHT TABLE (IMPERIAL AND METRIC) *(cont.)*

STEEL JOISTS (SLH SERIES)

Designation	Imperial		Metric	
	Depth (in)	Weight (lb/ft)	Depth (mm)	Weight (kg/m)
104SLH18	104	59	2642	88
104SLH19	104	67	2642	100
104SLH20	104	75	2642	112
104SLH21	104	90	2642	134
104SLH22	104	104	2642	155
104SLH23	104	109	2642	162
112SLH19	112	67	2845	100
112SLH20	112	76	2845	113
112SLH21	112	91	2845	136
112SLH22	112	104	2845	155
112SLH23	112	110	2845	164
112SLH24	112	131	2845	195
120SLH20	120	77	3048	115
120SLH21	120	92	3048	137
120SLH22	120	104	3048	155
120SLH23	120	111	3048	165
120SLH24	120	132	3048	197
120SLH25	120	152	3048	226

STRUCTURAL STEEL WEIGHT TABLE (IMPERIAL AND METRIC) *(cont.)*

W (WIDE FLANGE SHAPE)

Imperial		Metric	
Designation	Weight (lb/ft)	Designation	Weight (kg/m)
W 44 × 335	335	W 1100 × 499	489.0
W 44 × 290	290	W 1100 × 433	423.0
W 44 × 262	262	W 1100 × 390	382.0
W 44 × 230	230	W 1100 × 343	336.0
W 40 × 593	593	W 1000 × 883	865.0
W 40 × 503	503	W 1000 × 748	734.0
W 40 × 431	431	W 1000 × 642	629.0
W 40 × 397	397	W 1000 × 591	579.0
W 40 × 372	372	W 1000 × 554	543.0
W 40 × 362	362	W 1000 × 539	528.0
W 40 × 324	324	W 1000 × 483	473.0
W 40 × 297	297	W 1000 × 443	433.0
W 40 × 277	277	W 1000 × 412	404.0
W 40 × 249	249	W 1000 × 371	363.0
W 40 × 215	215	W 1000 × 321	314.0
W 40 × 199	199	W 1000 × 296	290.0
W 40 × 392	392	W 1000 × 584	572.0
W 40 × 331	331	W 1000 × 494	483.0
W 40 × 327	327	W 1000 × 486	477.0
W 40 × 278	278	W 1000 × 415	406.0
W 40 × 264	264	W 1000 × 393	385.0
W 40 × 235	235	W 1000 × 350	343.0
W 40 × 211	211	W 1000 × 314	308.0
W 40 × 183	183	W 1000 × 272	267.0
W 40 × 167	167	W 1000 × 249	244.0
W 40 × 149	149	W 1000 × 222	217.0
W 36 × 798	798	W 920 × 1188	1160.0

STRUCTURAL STEEL WEIGHT TABLE (IMPERIAL AND METRIC) (cont.)

W (WIDE FLANGE SHAPE)

Imperial		Metric	
Designation	Weight (lb/ft)	Designation	Weight (kg/m)
W 36 × 650	650	W 920 × 967	949.0
W 36 × 527	527	W 920 × 784	769.0
W 36 × 439	439	W 920 × 653	641.0
W 36 × 393	393	W 920 × 585	574.0
W 36 × 359	359	W 920 × 534	524.0
W 36 × 328	328	W 920 × 488	479.0
W 36 × 300	300	W 920 × 446	438.0
W 36 × 280	280	W 920 × 417	409.0
W 36 × 260	260	W 920 × 387	379.0
W 36 × 245	245	W 920 × 365	358.0
W 36 × 230	230	W 920 × 342	336.0
W 36 × 256	256	W 920 × 381	374.0
W 36 × 232	232	W 920 × 345	339.0
W 36 × 210	210	W 920 × 313	306.0
W 36 × 194	194	W 920 × 289	283.0
W 36 × 182	182	W 920 × 271	266.0
W 36 × 170	170	W 920 × 253	248.0
W 36 × 160	160	W 920 × 238	233.0
W 36 × 150	150	W 920 × 223	219.0
W 36 × 135	135	W 920 × 201	197.0
W 33 × 387	387	W 840 × 576	565.0
W 33 × 354	354	W 840 × 527	517.0
W 33 × 318	318	W 840 × 473	464.0
W 33 × 291	291	W 840 × 433	425.0
W 33 × 263	263	W 840 × 392	384.0
W 33 × 241	241	W 840 × 359	352.0
W 33 × 221	221	W 840 × 329	323.0

STRUCTURAL STEEL WEIGHT TABLE (IMPERIAL AND METRIC) *(cont.)*

W (WIDE FLANGE SHAPE)

Imperial		Metric	
Designation	Weight (lb/ft)	Designation	Weight (kg/m)
W 33 × 201	201	W 840 × 299	293.0
W 33 × 169	169	W 840 × 251	247.0
W 33 × 152	152	W 840 × 226	222.0
W 33 × 141	141	W 840 × 210	206.0
W 33 × 130	130	W 840 × 193	190.0
W 33 × 118	118	W 840 × 176	172.0
W 30 × 391	391	W 760 × 582	571.0
W 30 × 357	357	W 760 × 531	521.0
W 30 × 326	326	W 760 × 484	476.0
W 30 × 292	292	W 760 × 434	426.0
W 30 × 261	261	W 760 × 389	381.0
W 30 × 235	235	W 760 × 350	343.0
W 30 × 211	211	W 760 × 314	308.0
W 30 × 191	191	W 760 × 284	279.0
W 30 × 173	173	W 760 × 257	252.0
W 30 × 148	148	W 760 × 220	216.0
W 30 × 132	132	W 760 × 196	193.0
W 30 × 124	124	W 760 × 185	181.0
W 30 × 116	116	W 760 × 173	169.0
W 30 × 108	108	W 760 × 161	158.0
W 30 × 99	99	W 760 × 147	144.0
W 30 × 90	90	W 760 × 134	131.0
W 27 × 539	539	W 690 × 802	787.0
W 27 × 368	368	W 690 × 548	537.0
W 27 × 336	336	W 690 × 500	490.0
W 27 × 307	307	W 690 × 457	448.0
W 27 × 281	281	W 690 × 419	410.0

STRUCTURAL STEEL WEIGHT TABLE
(IMPERIAL AND METRIC) *(cont.)*

W (WIDE FLANGE SHAPE)

Imperial		Metric	
Designation	**Weight (lb/ft)**	**Designation**	**Weight (kg/m)**
W 27 × 258	258	W 690 × 384	377.0
W 27 × 235	235	W 690 × 350	343.0
W 27 × 217	217	W 690 × 323	317.0
W 27 × 194	194	W 690 × 289	283.0
W 27 × 178	178	W 690 × 265	260.0
W 27 × 161	161	W 690 × 240	235.0
W 27 × 146	146	W 690 × 217	213.0
W 27 × 129	129	W 690 × 192	188.0
W 27 × 114	114	W 690 × 170	166.0
W 27 × 102	102	W 690 × 152	149.0
W 27 × 94	94	W 690 × 140	137.0
W 27 × 84	84	W 690 × 125	123.0
W 24 × 370	370	W 610 × 551	540.0
W 24 × 335	335	W 610 × 498	489.0
W 24 × 306	306	W 610 × 455	447.0
W 24 × 279	279	W 610 × 415	407.0
W 24 × 250	250	W 610 × 372	365.0
W 24 × 229	229	W 610 × 341	334.0
W 24 × 207	207	W 610 × 307	302.0
W 24 × 192	192	W 610 × 285	280.0
W 24 × 176	176	W 610 × 262	257.0
W 24 × 162	162	W 610 × 241	236.0
W 24 × 146	146	W 610 × 217	213.0
W 24 × 131	131	W 610 × 195	191.0
W 24 × 117	117	W 610 × 174	171.0
W 24 × 104	104	W 610 × 155	152.0
W 24 × 103	103	W 610 × 153	150.0

W (WIDE FLANGE SHAPE)			
Imperial		Metric	
Designation	Weight (lb/ft)	Designation	Weight (kg/m)
W 24 × 94	94	W 610 × 140	137.0
W 24 × 84	84	W 610 × 125	123.0
W 24 × 76	76	W 610 × 113	111.0
W 24 × 68	68	W 610 × 101	99.2
W 24 × 62	62	W 610 × 92	90.5
W 24 × 55	55	W 610 × 82	80.3
W 21 × 201	201	W 530 × 300	293.0
W 21 × 182	182	W 530 × 272	266.0
W 21 × 166	166	W 530 × 248	242.0
W 21 × 147	147	W 530 × 219	215.0
W 21 × 132	132	W 530 × 196	193.0
W 21 × 122	122	W 530 × 182	178.0
W 21 × 111	111	W 530 × 165	162.0
W 21 × 101	101	W 530 × 150	147.0
W 21 × 93	93	W 530 × 138	136.0
W 21 × 83	83	W 530 × 123	121.0
W 21 × 73	73	W 530 × 109	107.0
W 21 × 68	68	W 530 × 101	99.2
W 21 × 62	62	W 530 × 92	90.5
W 21 × 55	55	W 530 × 82	80.3
W 21 × 48	48	W 530 × 72	70.0
W 21 × 57	57	W 530 × 85	83.2
W 21 × 50	50	W 530 × 74	73.0
W 21 × 44	44	W 530 × 66	64.2
W 18 × 175	175	W 460 × 260	255.0
W 18 × 158	158	W 460 × 235	231.0
W 18 × 143	143	W 460 × 213	209.0

STRUCTURAL STEEL WEIGHT TABLE (IMPERIAL AND METRIC) (cont.)

W (WIDE FLANGE SHAPE)

Imperial		Metric	
Designation	Weight (lb/ft)	Designation	Weight (kg/m)
W 18 × 130	130	W 460 × 193	190.0
W 18 × 119	119	W 460 × 177	174.0
W 18 × 106	106	W 460 × 158	155.0
W 18 × 97	97	W 460 × 144	142.0
W 18 × 86	86	W 460 × 128	126.0
W 18 × 76	76	W 460 × 113	111.0
W 18 × 71	71	W 460 × 106	104.0
W 18 × 65	65	W 460 × 97	94.9
W 18 × 60	60	W 460 × 89	87.6
W 18 × 55	55	W 460 × 82	80.3
W 18 × 50	50	W 460 × 74	73.0
W 18 × 46	46	W 460 × 68	67.1
W 18 × 40	40	W 460 × 60	58.4
W 18 × 35	35	W 460 × 52	51.1
W 16 × 100	100	W 410 × 149	146.0
W 16 × 89	89	W 410 × 132	130.0
W 16 × 77	77	W 410 × 114	112.0
W 16 × 67	67	W 410 × 100	97.8
W 16 × 57	57	W 410 × 85	83.2
W 16 × 50	50	W 410 × 75	73.0
W 16 × 45	45	W 410 × 67	65.7
W 16 × 40	40	W 410 × 60	58.4
W 16 × 36	36	W 410 × 53	52.5
W 16 × 31	31	W 410 × 46.1	45.2
W 16 × 26	26	W 410 × 38.8	37.9
W 14 × 808	808	W 360 × 1202	1180.0
W 14 × 730	730	W 360 × 1086	1070.0

STRUCTURAL STEEL WEIGHT TABLE (IMPERIAL AND METRIC) *(cont.)*

W (WIDE FLANGE SHAPE)

Imperial		Metric	
Designation	**Weight (lb/ft)**	**Designation**	**Weight (kg/m)**
W 14 × 665	665	W 360 × 990	970.0
W 14 × 605	605	W 360 × 900	883.0
W 14 × 550	550	W 360 × 818	803.0
W 14 × 500	500	W 360 × 744	730.0
W 14 × 455	455	W 360 × 677	664.0
W 14 × 426	426	W 360 × 634	622.0
W 14 × 398	398	W 360 × 592	581.0
W 14 × 370	370	W 360 × 551	540.0
W 14 × 342	342	W 360 × 509	499.0
W 14 × 311	311	W 360 × 463	454.0
W 14 × 283	283	W 360 × 421	413.0
W 14 × 257	257	W 360 × 382	375.0
W 14 × 233	233	W 360 × 347	340.0
W 14 × 211	211	W 360 × 314	308.0
W 14 × 193	193	W 360 × 287	282.0
W 14 × 176	176	W 360 × 262	257.0
W 14 × 159	159	W 360 × 237	232.0
W 14 × 145	145	W 360 × 216	212.0
W 14 × 132	132	W 360 × 196	193.0
W 14 × 120	120	W 360 × 179	175.0
W 14 × 109	109	W 360 × 162	159.0
W 14 × 99	99	W 360 × 147	144.0
W 14 × 90	90	W 360 × 134	131.0
W 14 × 82	82	W 360 × 122	120.0
W 14 × 74	74	W 360 × 110	108.0
W 14 × 68	68	W 360 × 101	99.2
W 14 × 61	61	W 360 × 91	89.0

STRUCTURAL STEEL WEIGHT TABLE (IMPERIAL AND METRIC) *(cont.)*

W (WIDE FLANGE SHAPE)

Imperial		Metric	
Designation	**Weight (lb/ft)**	**Designation**	**Weight (kg/m)**
W 14 × 53	53	W 360 × 79	77.3
W 14 × 48	48	W 360 × 72	70.0
W 14 × 43	43	W 360 × 64	62.8
W 14 × 38	38	W 360 × 57.8	55.5
W 14 × 34	34	W 360 × 51	49.6
W 14 × 30	30	W 360 × 44	43.8
W 14 × 26	26	W 360 × 39	37.9
W 14 × 22	22	W 360 × 32.9	32.1
W 12 × 336	336	W 310 × 500	490.0
W 12 × 305	305	W 310 × 454	445.0
W 12 × 279	279	W 310 × 415	407.0
W 12 × 252	252	W 310 × 375	368.0
W 12 × 230	230	W 310 × 342	336.0
W 12 × 210	210	W 310 × 313	306.0
W 12 × 190	190	W 310 × 283	277.0
W 12 × 170	170	W 310 × 253	248.0
W 12 × 152	152	W 310 × 226	222.0
W 12 × 136	136	W 310 × 202	198.0
W 12 × 120	120	W 310 × 179	175.0
W 12 × 106	106	W 310 × 158	155.0
W 12 × 96	96	W 310 × 143	140.0
W 12 × 87	87	W 310 × 129	127.0
W 12 × 79	79	W 310 × 117	115.0
W 12 × 72	72	W 310 × 107	105.0
W 12 × 65	65	W 310 × 97	94.9
W 12 × 58	58	W 310 × 86	84.6
W 12 × 53	53	W 310 × 79	77.3

STRUCTURAL STEEL WEIGHT TABLE
(IMPERIAL AND METRIC) *(cont.)*

W (WIDE FLANGE SHAPE)

Imperial		Metric	
Designation	Weight (lb/ft)	Designation	Weight (kg/m)
W 12 × 50	50	W 310 × 74	73.0
W 12 × 45	45	W 310 × 67	65.7
W 12 × 40	40	W 310 × 60	58.4
W 12 × 35	35	W 310 × 52	51.1
W 12 × 30	30	W 310 × 44.5	43.8
W 12 × 26	26	W 310 × 38.7	37.9
W 12 × 22	22	W 310 × 32.7	32.1
W 12 × 19	19	W 310 × 28.3	27.7
W 12 × 16	16	W 310 × 23.8	23.3
W 12 × 14	14	W 310 × 21	20.4
W 10 × 112	112	W 250 × 167	163.0
W 10 × 100	100	W 250 × 149	146.0
W 10 × 88	88	W 250 × 131	128.0
W 10 × 77	77	W 250 × 115	112.0
W 10 × 68	68	W 250 × 101	99.2
W 10 × 60	60	W 250 × 89	87.6
W 10 × 54	54	W 250 × 80	78.8
W 10 × 49	49	W 250 × 73	71.5
W 10 × 45	45	W 250 × 67	65.7
W 10 × 39	39	W 250 × 58	56.9
W 10 × 33	33	W 250 × 49.1	48.2
W 10 × 30	30	W 250 × 44.8	43.8
W 10 × 26	26	W 250 × 38.5	37.9
W 10 × 22	22	W 250 × 32.7	32.1
W 10 × 19	19	W 250 × 28.4	27.7
W 10 × 17	17	W 250 × 25.3	24.8
W 10 × 15	15	W 250 × 22.3	21.9

STRUCTURAL STEEL WEIGHT TABLE (IMPERIAL AND METRIC) *(cont.)*

W (WIDE FLANGE SHAPE)

Imperial		Metric	
Designation	Weight (lb/ft)	Designation	Weight (kg/m)
W 10 × 12	12	W 250 × 17.9	17.5
W 8 × 67	67	W 200 × 100	97.8
W 8 × 58	58	W 200 × 86	84.6
W 8 × 48	48	W 200 × 71	70.0
W 8 × 40	40	W 200 × 59	58.4
W 8 × 35	35	W 200 × 52	51.1
W 8 × 31	31	W 200 × 46.1	45.2
W 8 × 28	28	W 200 × 41.7	40.9
W 8 × 24	24	W 200 × 35.9	35.0
W 8 × 21	21	W 200 × 31.3	30.6
W 8 × 18	18	W 200 × 26.6	26.3
W 8 × 15	15	W 200 × 22.5	21.9
W 8 × 13	13	W 200 × 19.3	19.0
W 8 × 10	10	W 200 × 15	14.6
W 6 × 25	25	W 150 × 37.1	36.5
W 6 × 20	20	W 150 × 29.8	29.2
W 6 × 15	15	W 150 × 22.5	21.9
W 6 × 16	16	W 150 × 24	23.3
W 6 × 12	12	W 150 × 18	17.5
W 6 × 9	9	W 150 × 13.5	13.1
W 6 × 8.5	8.5	W 150 × 13	12.4
W 5 × 19	19	W 130 × 28.1	27.7
W 5 × 16	16	W 130 × 23.8	23.3
W 4 × 13	13	W 100 × 19.3	19.0

STRUCTURAL STEEL WEIGHT TABLE (IMPERIAL AND METRIC) *(cont.)*

M (MISCELLANEOUS BEAMS)

Imperial		Metric	
Designation	Weight (lb/ft)	Designation	Weight (kg/m)
M 12 x 11.8	11.8	M 310 x 17.6	17.2
M 12 x 10.8	10.8	M 310 x 16.1	15.8
M 12 x 10	10	M 310 x 14.9	14.6
M 10 x 9	9	M 250 x 13.4	13.1
M 10 x 8	8	M 250 x 11.9	11.7
M 10 x 7.5	7.5	M 250 x 11.2	10.9
M 8 x 6.5	6.5	M 200 x 9.7	9.5
M 8 x 6.2	6.2	M 200 x 9.2	9.1
M 6 x 4.4	4.4	M 150 x 6.6	6.4
M 6 x 3.7	3.7	M 150 x 5.5	5.4
M 5 x 18.9	18.9	M 130 x 28.1	27.6
M 4 x 6	6	M 100 x 8.9	8.8

STRUCTURAL STEEL WEIGHT TABLE (IMPERIAL AND METRIC) *(cont.)*

S (AMERICAN STANDARD BEAMS)

Imperial		Metric	
Designation	Weight (lb/ft)	Designation	Weight (kg/m)
S 24 × 121	121	S 610 × 180	177.0
S 24 × 106	106	S 610 × 158	155.0
S 24 × 100	100	S 610 × 149	146.0
S 24 × 90	90	S 610 × 134	131.0
S 24 × 80	80	S 610 × 119	117.0
S 20 × 96	96	S 510 × 143	140.0
S 20 × 86	86	S 510 × 128	126.0
S 20 × 75	75	S 510 × 112	109.0
S 20 × 66	66	S 510 × 98.2	96.3
S 18 × 70	70	S 460 × 104	102.0
S 18 × 54.7	54.7	S 460 × 81.4	79.8
S 15 × 50	50	S 380 × 74	73.0
S 15 × 42.9	42.9	S 380 × 64	62.6
S 12 × 50	50	S 310 × 74	73.0
S 12 × 40.8	40.8	S 310 × 60.7	59.5
S 12 × 35	35	S 310 × 52	51.1
S 12 × 31.8	31.8	S 310 × 47.3	46.4
S 10 × 35	35	S 250 × 52	51.1
S 10 × 25.4	25.4	S 250 × 37.8	37.1
S 8 × 23	23	S 200 × 34	33.6
S 8 × 18.4	18.4	S 200 × 27.4	26.9
S 6 × 17.25	17.25	S 150 × 25.7	25.2
S 6 × 12.5	12.5	S 150 × 18.6	18.2
S 5 × 10	10	S 130 × 15	14.6
S 4 × 9.5	9.5	S 100 × 14.1	13.9
S 4 × 7.7	7.7	S 100 × 11.5	11.2
S 3 × 7.5	7.5	S 75 × 11.2	10.9
S 3 × 5.7	5.7	S 75 × 8.5	8.3

STRUCTURAL STEEL WEIGHT TABLE (IMPERIAL AND METRIC) *(cont.)*

HP (BEARING PILES)

Imperial		Metric	
Designation	Weight (lb/ft)	Designation	Weight (kg/m)
HP 14 × 117	117	HP 360 × 174	171.0
HP 14 × 102	102	HP 360 × 152	149.0
HP 14 × 89	89	HP 360 × 132	130.0
HP 14 × 73	73	HP 360 × 108	107.0
HP 12 × 84	84	HP 310 × 125	123.0
HP 12 × 74	74	HP 310 × 110	108.0
HP 12 × 63	63	HP 310 × 93	91.9
HP 12 × 53	53	HP 310 × 79	77.3
HP 10 × 57	57	HP 250 × 85	83.2
HP 10 × 42	42	HP 250 × 62	61.3
HP 8 × 36	36	HP 200 × 53	52.5

STRUCTURAL STEEL WEIGHT TABLE (IMPERIAL AND METRIC) *(cont.)*

C (AMERICAN STANDARD CHANNELS)

Imperial		Metric	
Designation	**Weight (lb/ft)**	**Designation**	**Weight (kg/m)**
C 15 × 50	50	C 380 × 74	73.0
C 15 × 40	40	C 380 × 60	58.4
C 15 × 33.9	33.9	C 380 × 50.4	49.5
C 12 × 30	30	C 310 × 45	43.8
C 12 × 25	25	C 310 × 37	36.5
C 12 × 20.7	20.7	C 310 × 30.8	30.2
C 10 × 30	30	C 250 × 45	43.8
C 10 × 25	25	C 250 × 37	36.5
C 10 × 20	20	C 250 × 30	29.2
C 10 × 15.3	15.3	C 250 × 22.8	22.3
C 9 × 20	20	C 230 × 30	29.2
C 9 × 15	15	C 230 × 22	21.9
C 9 × 13.4	13.4	C 230 × 19.9	19.6
C 8 × 18.75	18.75	C 200 × 27.9	27.4
C 8 × 13.75	13.75	C 200 × 20.5	20.1
C 8 × 11.5	11.5	C 200 × 17.1	16.8
C 7 × 14.75	14.75	C 180 × 22	21.5
C 7 × 12.25	12.25	C 180 × 18.2	17.9
C 7 × 9.8	9.8	C 180 × 14.6	14.3
C 6 × 13	13	C 150 × 19.3	19.0
C 6 × 10.5	10.5	C 150 × 15.6	15.3
C 6 × 8.2	8.2	C 150 × 12.2	12.0
C 5 × 9	9	C 130 × 13	13.1
C 5 × 6.7	6.7	C 130 × 10.4	9.8
C 4 × 7.25	7.25	C 100 × 10.8	10.6
C 4 × 5.4	5.4	C 100 × 8	7.9
C 4 × 4.5	4.5	C 100 × 6.7	6.6
C 3 × 6	6	C 75 × 8.9	8.8
C 3 × 5	5	C 75 × 7.4	7.3
C 3 × 4.1	4.1	C 75 × 6.1	6.0
C 3 × 3.5	3.5	C 75 × 5.2	5.1

STRUCTURAL STEEL WEIGHT TABLE (IMPERIAL AND METRIC) *(cont.)*

MC (MISCELLANEOUS CHANNELS)

Imperial		Metric	
Designation	Weight (lb/ft)	Designation	Weight (kg/m)
MC 18 × 58	58	MC 460 × 86	84.6
MC 18 × 51.9	51.9	MC 460 × 77.2	75.7
MC 18 × 45.8	45.8	MC 460 × 68.2	66.8
MC 18 × 42.7	42.7	MC 460 × 63.5	62.3
MC 13 × 50	50	MC 330 × 74	73.0
MC 13 × 40	40	MC 330 × 60	58.4
MC 13 × 35	35	MC 330 × 52	51.1
MC 13 × 31.8	31.8	MC 330 × 47.3	46.4
MC 12 × 50	50	MC 310 × 74	73.0
MC 12 × 45	45	MC 310 × 67	65.7
MC 12 × 40	40	MC 310 × 60	58.4
MC 12 × 35	35	MC 310 × 52	51.1
MC 12 × 31	31	MC 310 × 46	45.2
MC 12 × 10.6	10.6	MC 310 × 15.8	15.5
MC 10 × 41.1	41.1	MC 250 × 61.2	60.0
MC 10 × 33.6	33.6	MC 250 × 50	49.0
MC 10 × 28.5	28.5	MC 250 × 42.4	41.6
MC 10 × 25	25	MC 250 × 37	36.5
MC 10 × 22	22	MC 250 × 33	32.1
MC 10 × 8.4	8.4	MC 250 × 12.5	12.3
MC 9 × 25.4	25.4	MC 230 × 37.8	37.1
MC 9 × 23.9	23.9	MC 230 × 35.6	34.9
MC 8 × 22.8	22.8	MC 200 × 33.9	33.3
MC 8 × 21.4	21.4	MC 200 × 31.8	31.2
MC 8 × 20	20	MC 200 × 29.8	29.2
MC 8 × 18.7	18.7	MC 200 × 27.8	27.3
MC 8 × 8.5	8.5	MC 200 × 12.6	12.4
MC 7 × 22.7	22.7	MC 180 × 33.8	33.1
MC 7 × 19.1	19.1	MC 180 × 28.4	27.9
MC 6 × 18	18	MC 150 × 26.8	26.3
MC 6 × 15.3	15.3	MC 150 × 22.8	22.3
MC 6 × 16.3	16.3	MC 150 × 24.3	23.8
MC 6 × 15.1	15.1	MC 150 × 22.5	22.0
MC 6 × 12	12	MC 150 × 17.9	17.5

STRUCTURAL STEEL WEIGHT TABLE
(IMPERIAL AND METRIC) *(cont.)*

WT (WIDE FLANGE TEES)

Imperial		Metric	
Designation	**Weight (lb/ft)**	**Designation**	**Weight (kg/m)**
WT 22 × 167.5	167.5	WT 550 × 249.5	244.0
WT 22 × 145	145	WT 550 × 216.5	212.0
WT 22 × 131	131	WT 550 × 195	191.0
WT 22 × 115	115	WT 550 × 171.5	168.0
WT 20 × 296.5	296.5	WT 500 × 441.5	433.0
WT 20 × 251.5	251.5	WT 500 × 374	367.0
WT 20 × 215.5	215.5	WT 500 × 321	314.0
WT 20 × 198.5	198.5	WT 500 × 295.5	290.0
WT 20 × 186	186	WT 500 × 277	271.0
WT 20 × 181	181	WT 500 × 269.5	264.0
WT 20 × 162	162	WT 500 × 241.5	236.0
WT 20 × 148.5	148.5	WT 500 × 221.5	217.0
WT 20 × 138.5	138.5	WT 500 × 206	202.0
WT 20 × 124.5	124.5	WT 500 × 185.5	182.0
WT 20 × 107.5	107.5	WT 500 × 160.5	157.0
WT 20 × 99.5	99.5	WT 500 × 148	145.0
WT 20 × 196	196	WT 500 × 292	286.0
WT 20 × 165.5	165.5	WT 500 × 247	242.0
WT 20 × 163.5	163.5	WT 500 × 243	239.0
WT 20 × 139	139	WT 500 × 207.5	203.0
WT 20 × 132	132	WT 500 × 196.5	193.0
WT 20 × 117.5	117.5	WT 500 × 175	171.0
WT 20 × 105.5	105.5	WT 500 × 157	154.0
WT 20 × 91.5	91.5	WT 500 × 136	134.0
WT 20 × 83.5	83.5	WT 500 × 124.5	122.0
WT 20 × 74.5	74.5	WT 500 × 111	109.0
WT 18 × 399	399	WT 460 × 594	582.0
WT 18 × 325	325	WT 460 × 483.5	474.0
WT 18 × 263.5	263.5	WT 460 × 392	385.0
WT 18 × 219.5	219.5	WT 460 × 326.5	320.0
WT 18 × 196.5	196.5	WT 460 × 292.5	287.0
WT 18 × 179.5	179.5	WT 460 × 267	262.0
WT 18 × 164	164	WT 460 × 244	239.0
WT 18 × 150	150	WT 460 × 223	219.0
WT 18 × 140	140	WT 460 × 208.5	204.0

STRUCTURAL STEEL WEIGHT TABLE
(IMPERIAL AND METRIC) (cont.)

WT (WIDE FLANGE TEES)

Imperial		Metric	
Designation	Weight (lb/ft)	Designation	Weight (kg/m)
WT 18 × 130	130	WT 460 × 193.5	190.0
WT 18 × 122.5	122.5	WT 460 × 182.5	179.0
WT 18 × 115	115	WT 460 × 171	168.0
WT 18 × 128	128	WT 460 × 190.5	187.0
WT 18 × 116	116	WT 460 × 172.5	169.0
WT 18 × 105	105	WT 460 × 156.5	153.0
WT 18 × 97	97	WT 460 × 144.5	142.0
WT 18 × 91	91	WT 460 × 135.5	133.0
WT 18 × 85	85	WT 460 × 126.5	124.0
WT 18 × 80	80	WT 460 × 119	117.0
WT 18 × 75	75	WT 460 × 111.5	109.0
WT 18 × 67.5	67.5	WT 460 × 100.5	98.5
WT 16.5 × 193.5	193.5	WT 420 × 288	282.0
WT 16.5 × 177	177	WT 420 × 249	258.0
WT 16.5 × 159	159	WT 420 × 236.5	232.0
WT 16.5 × 145.5	145.5	WT 420 × 216.5	212.0
WT 16.5 × 131.5	131.5	WT 420 × 196	192.0
WT 16.5 × 120.5	120.5	WT 420 × 179.5	176.0
WT 16.5 × 110.5	110.5	WT 420 × 164.5	161.0
WT 16.5 × 100.5	100.5	WT 420 × 149.5	147.0
WT 16.5 × 84.5	84.5	WT 420 × 125.5	123.0
WT 16.5 × 76	76	WT 420 × 113	111.0
WT 16.5 × 70.5	70.5	WT 420 × 105	103.0
WT 16.5 × 65	65	WT 420 × 96.5	94.9
WT 16.5 × 59	59	WT 420 × 88	86.1
WT 15 × 195.5	195.5	WT 380 × 291	285.0
WT 15 × 178.5	178.5	WT 380 × 265.5	260.0
WT 15 × 163	163	WT 380 × 242	238.0
WT 15 × 146	146	WT 380 × 217	213.0
WT 15 × 130.5	130.5	WT 380 × 194.5	190.0
WT 15 × 117.5	117.5	WT 380 × 175	171.0
WT 15 × 105.5	105.5	WT 380 × 157	154.0
WT 15 × 95.5	95.5	WT 380 × 142	139.0
WT 15 × 86.5	86.5	WT 380 × 128.5	126.0
WT 15 × 74	74	WT 380 × 110	108.0

STRUCTURAL STEEL WEIGHT TABLE
(IMPERIAL AND METRIC) (cont.)

WT (WIDE FLANGE TEES)

Imperial		Metric	
Designation	**Weight (lb/ft)**	**Designation**	**Weight (kg/m)**
WT 15 × 66	66	WT 380 × 98	96.3
WT 15 × 62	62	WT 380 × 92.5	90.5
WT 15 × 58	58	WT 380 × 86.5	84.6
WT 15 × 54	54	WT 380 × 80.5	78.8
WT 15 × 49.5	49.5	WT 380 × 73.5	72.2
WT 15 × 45	45	WT 380 × 67	65.7
WT 13.5 × 269.5	269.5	WT 345 × 401	393.0
WT 13.5 × 184	184	WT 345 × 274	269.0
WT 13.5 × 168	168	WT 345 × 250	245.0
WT 13.5 × 153.5	153.5	WT 345 × 228.5	224.0
WT 13.5 × 140.5	140.5	WT 345 × 209.5	205.0
WT 13.5 × 129	129	WT 345 × 192	188.0
WT 13.5 × 117.5	117.5	WT 345 × 175	171.0
WT 13.5 × 108.5	108.5	WT 345 × 161.5	158.0
WT 13.5 × 97	97	WT 345 × 144.5	142.0
WT 13.5 × 89	89	WT 345 × 132.5	130.0
WT 13.5 × 80.5	80.5	WT 345 × 120	117.0
WT 13.5 × 73	73	WT 345 × 108.5	107.0
WT 13.5 × 64.5	64.5	WT 345 × 96	94.1
WT 13.5 × 57	57	WT 345 × 85	83.2
WT 13.5 × 51	51	WT 345 × 76	74.4
WT 13.5 × 47	47	WT 345 × 70	68.6
WT 13.5 × 42	42	WT 345 × 62.5	61.3
WT 12 × 185	185	WT 305 × 275.5	270.0
WT 12 × 167.5	167.5	WT 305 × 249	244.0
WT 12 × 153	153	WT 305 × 227.5	223.0
WT 12 × 139.5	139.5	WT 305 × 207.5	204.0
WT 12 × 125	125	WT 305 × 186	182.0
WT 12 × 114.5	114.5	WT 305 × 170.5	167.0
WT 12 × 103.5	103.5	WT 305 × 153.5	151.0
WT 12 × 96	96	WT 305 × 142.5	140.0
WT 12 × 88	88	WT 305 × 131	128.0
WT 12 × 81	81	WT 305 × 120.5	118.0
WT 12 × 73	73	WT 305 × 108.5	107.0
WT 12 × 65.5	65.5	WT 305 × 97.5	95.6

STRUCTURAL STEEL WEIGHT TABLE
(IMPERIAL AND METRIC) *(cont.)*

WT (WIDE FLANGE TEES)

Imperial		Metric	
Designation	**Weight (lb/ft)**	**Designation**	**Weight (kg/m)**
WT 12 × 58.5	58.5	WT 305 × 87	85.4
WT 12 × 52	52	WT 305 × 77.5	75.9
WT 12 × 51.5	51.5	WT 305 × 76.5	75.2
WT 12 × 47	47	WT 305 × 70	68.6
WT 12 × 42	42	WT 305 × 62.5	61.3
WT 12 × 38	38	WT 305 × 56.5	55.5
WT 12 × 34	34	WT 305 × 50.5	49.6
WT 12 × 31	31	WT 305 × 46	45.2
WT 12 × 27.5	27.5	WT 305 × 41	40.1
WT 10.5 × 100.5	100.5	WT 265 × 150	147.0
WT 10.5 × 91	91	WT 265 × 136	133.0
WT 10.5 × 83	83	WT 265 × 124	121.0
WT 10.5 × 73.5	73.5	WT 265 × 109.5	107.0
WT 10.5 × 66	66	WT 265 × 98	96.3
WT 10.5 × 61	61	WT 265 × 91	89.0
WT 10.5 × 55.5	55.5	WT 265 × 82.5	81.0
WT 10.5 × 50.5	50.5	WT 265 × 75	73.7
WT 10.5 × 46.5	46.5	WT 265 × 69	67.9
WT 10.5 × 41.5	41.5	WT 265 × 61.5	60.6
WT 10.5 × 36.5	36.5	WT 265 × 54.5	53.3
WT 10.5 × 34	34	WT 265 × 50.5	49.6
WT 10.5 × 31	31	WT 265 × 46	45.2
WT 10.5 × 27.5	27.5	WT 265 × 41	40.1
WT 10.5 × 24	24	WT 265 × 36	35.0
WT 10.5 × 28.5	28.5	WT 265 × 42.5	41.6
WT 10.5 × 25	25	WT 265 × 37	36.5
WT 10.5 × 22	22	WT 265 × 33	32.1
WT 9 × 87.5	87.5	WT 230 × 130	128.0
WT 9 × 79	79	WT 230 × 117.5	115.0
WT 9 × 71.5	71.5	WT 230 × 106.5	104.0
WT 9 × 65	65	WT 230 × 96.5	94.9
WT 9 × 59.5	59.5	WT 230 × 88.5	86.8
WT 9 × 53	53	WT 230 × 79	77.3
WT 9 × 48.5	48.5	WT 230 × 72	70.8
WT 9 × 43	43	WT 230 × 64	62.8

STRUCTURAL STEEL WEIGHT TABLE
(IMPERIAL AND METRIC) *(cont.)*
WT (WIDE FLANGE TEES)

Imperial		Metric	
Designation	**Weight (lb/ft)**	**Designation**	**Weight (kg/m)**
WT 9 × 38	38	WT 230 × 56.5	55.5
WT 9 × 35.5	35.5	WT 230 × 53	51.8
WT 9 × 32.5	32.5	WT 230 × 48.5	47.4
WT 9 × 30	30	WT 230 × 44.5	43.8
WT 9 × 27.5	27.5	WT 230 × 41	40.1
WT 9 × 25	25	WT 230 × 37	36.5
WT 9 × 23	23	WT 230 × 34	33.6
WT 9 × 20	20	WT 230 × 30	29.2
WT 9 × 17.5	17.5	WT 230 × 26	25.5
WT 8 × 50	50	WT 205 × 74.5	73.0
WT 8 × 44.5	44.5	WT 205 × 66	64.9
WT 8 × 38.5	38.5	WT 205 × 57	56.2
WT 8 × 33.5	33.5	WT 205 × 50	48.9
WT 8 × 28.5	28.5	WT 205 × 42.5	41.6
WT 8 × 25	25	WT 205 × 37.5	36.5
WT 8 × 22.5	22.5	WT 205 × 33.5	32.8
WT 8 × 20	20	WT 205 × 30	29.2
WT 8 × 18	18	WT 205 × 26.5	26.3
WT 8 × 15.5	15.5	WT 205 × 23.05	22.6
WT 8 × 13	13	WT 205 × 19.4	19.0
WT 7 × 404	404	WT 180 × 601	590.0
WT 7 × 365	365	WT 180 × 543	533.0
WT 7 × 332.5	332.5	WT 180 × 495	485.0
WT 7 × 302.5	302.5	WT 180 × 450	441.0
WT 7 × 275	275	WT 180 × 409	401.0
WT 7 × 250	250	WT 180 × 372	365.0
WT 7 × 227.5	227.5	WT 180 × 338.5	332.0
WT 7 × 213	213	WT 180 × 317	311.0
WT 7 × 199	199	WT 180 × 296	290.0
WT 7 × 185	185	WT 180 × 275.5	270.0
WT 7 × 171	171	WT 180 × 254.5	250.0
WT 7 × 155.5	155.5	WT 180 × 231.5	227.0
WT 7 × 141.5	141.5	WT 180 × 210.5	206.0
WT 7 × 128.5	128.5	WT 180 × 191	188.0
WT 7 × 116.5	116.5	WT 180 × 173.5	170.0

STRUCTURAL STEEL WEIGHT TABLE
(IMPERIAL AND METRIC) *(cont.)*

WT (WIDE FLANGE TEES)

Imperial		Metric	
Designation	**Weight (lb/ft)**	**Designation**	**Weight (kg/m)**
WT 7 × 105.5	105.5	WT 180 × 157	154.0
WT 7 × 96.5	96.5	WT 180 × 143.5	141.0
WT 7 × 88	88	WT 180 × 131	128.0
WT 7 × 79.5	79.5	WT 180 × 118.5	116.0
WT 7 × 72.5	72.5	WT 180 × 108	106.0
WT 7 × 66	66	WT 180 × 98	96.3
WT 7 × 60	60	WT 180 × 89.5	87.6
WT 7 × 54.5	54.5	WT 180 × 81	79.5
WT 7 × 49.5	49.5	WT 180 × 73.5	72.2
WT 7 × 45	45	WT 180 × 67	65.7
WT 7 × 41	41	WT 180 × 61	59.8
WT 7 × 37	37	WT 180 × 55	54.0
WT 7 × 34	34	WT 180 × 50.5	49.6
WT 7 × 30.5	30.5	WT 180 × 45.5	44.5
WT 7 × 26.5	26.5	WT 180 × 39.5	38.7
WT 7 × 24	24	WT 180 × 36	35.0
WT 7 × 21.5	21.5	WT 180 × 32	31.4
WT 7 × 19	19	WT 180 × 28.9	27.7
WT 7 × 17	17	WT 180 × 25.5	24.8
WT 7 × 15	15	WT 180 × 22	21.9
WT 7 × 13	13	WT 180 × 19.5	19.0
WT 7 × 11	11	WT 180 × 16.45	16.1
WT 6 × 168	168	WT 155 × 250	245.0
WT 6 × 152.5	152.5	WT 155 × 227	223.0
WT 6 × 139.5	139.5	WT 155 × 207.5	204.0
WT 6 × 126	126	WT 155 × 187.5	184.0
WT 6 × 115	115	WT 155 × 171	168.0
WT 6 × 105	105	WT 155 × 156.5	153.0
WT 6 × 95	95	WT 155 × 141.5	139.0
WT 6 × 85	85	WT 155 × 126.5	124.0
WT 6 × 76	76	WT 155 × 113	111.0
WT 6 × 68	68	WT 155 × 101	99.2
WT 6 × 60	60	WT 155 × 89.5	87.6
WT 6 × 53	53	WT 155 × 79	77.3
WT 6 × 48	48	WT 155 × 71.5	70.0

STRUCTURAL STEEL WEIGHT TABLE (IMPERIAL AND METRIC) (cont.)

WT (WIDE FLANGE TEES)

Imperial		Metric	
Designation	Weight (lb/ft)	Designation	Weight (kg/m)
WT 6 × 43.5	43.5	WT 155 × 64.5	63.5
WT 6 × 39.5	39.5	WT 155 × 58.5	57.6
WT 6 × 36	36	WT 155 × 53.5	52.5
WT 6 × 32.5	32.5	WT 155 × 48.5	47.4
WT 6 × 29	29	WT 155 × 43	42.3
WT 6 × 26.5	26.5	WT 155 × 39.5	38.7
WT 6 × 25	25	WT 155 × 37	36.5
WT 6 × 22.5	22.5	WT 155 × 33.5	32.8
WT 6 × 20	20	WT 155 × 30	29.2
WT 6 × 17.5	17.5	WT 155 × 26	25.5
WT 6 × 15	15	WT 155 × 22.25	21.9
WT 6 × 13	13	WT 155 × 19.35	19.0
WT 6 × 11	11	WT 155 × 16.35	16.1
WT 6 × 9.5	9.5	WT 155 × 14.15	13.9
WT 6 × 8	8	WT 155 × 11.9	11.7
WT 6 × 7	7	WT 155 × 10.5	10.2
WT 5 × 56	56	WT 125 × 83.5	81.7
WT 5 × 50	50	WT 125 × 74.5	73.0
WT 5 × 44	44	WT 125 × 65.5	64.2
WT 5 × 38.5	38.5	WT 125 × 57.5	56.2
WT 5 × 34	34	WT 125 × 50.5	49.6
WT 5 × 30	30	WT 125 × 44.5	43.8
WT 5 × 27	27	WT 125 × 40	39.4
WT 5 × 24.5	24.5	WT 125 × 36.5	35.8
WT 5 × 22.5	22.5	WT 125 × 33.5	32.8
WT 5 × 19.5	19.5	WT 125 × 29	28.5
WT 5 × 16.5	16.5	WT 125 × 24.55	24.1
WT 5 × 15	15	WT 125 × 22.4	21.9
WT 5 × 13	13	WT 125 × 19.25	19.0
WT 5 × 11	11	WT 125 × 16.35	16.1
WT 5 × 9.5	9.5	WT 125 × 14.2	13.9
WT 5 × 8.5	8.5	WT 125 × 12.65	12.4
WT 5 × 7.5	7.5	WT 125 × 11.15	10.9
WT 5 × 6	6	WT 125 × 8.95	8.8
WT 4 × 33.5	33.5	WT 100 × 50	48.9

STRUCTURAL STEEL WEIGHT TABLE
(IMPERIAL AND METRIC) *(cont.)*

WT (WIDE FLANGE TEES)			
Imperial		**Metric**	
Designation	**Weight (lb/ft)**	**Designation**	**Weight (kg/m)**
WT 4 × 29	29	WT 100 × 43	42.3
WT 4 × 24	24	WT 100 × 35.5	35.0
WT 4 × 20	20	WT 100 × 29.5	29.2
WT 4 × 17.5	17.5	WT 100 × 26	25.5
WT 4 × 15.5	15.5	WT 100 × 23.05	22.6
WT 4 × 14	14	WT 100 × 20.85	20.4
WT 4 × 12	12	WT 100 × 17.95	17.5
WT 4 × 10.5	10.5	WT 100 × 15.65	15.3
WT 4 × 9	9	WT 100 × 13.3	13.1
WT 4 × 7.5	7.5	WT 100 × 11.25	10.9
WT 4 × 6.5	6.5	WT 100 × 9.65	9.5
WT 4 × 5	5	WT 100 × 7.5	7.3
WT 3 × 12.5	12.5	WT 75 × 18.55	18.2
WT 3 × 10	10	WT 75 × 14.9	14.6
WT 3 × 7.5	7.5	WT 75 × 11.25	10.9
WT 3 × 8	8	WT 75 × 12	11.7
WT 3 × 6	6	WT 75 × 9	8.8
WT 3 × 4.5	4.5	WT 75 × 6.75	6.6
WT 3 × 4.25	4.25	WT 75 × 6.5	6.2
WT 2.5 × 9.5	9.5	WT 65 × 14.05	13.9
WT 2.5 × 8	8	WT 65 × 11.9	11.7
WT 2 × 6.5	6.5	WT 50 × 9.65	9.5
MT (Miscellaneous Tees)			
MT 6 × 5.9	5.9	MT 155 × 8.8	8.6
MT 6 × 5.4	5.4	MT 155 × 8.05	7.9
MT 6 × 5	5	MT 155 × 7.45	7.3
MT 5 × 4.5	4.5	MT 125 × 6.7	6.6
MT 5 × 4	4	MT 125 × 5.95	5.8
MT 5 × 3.75	3.75	MT 125 × 5.6	5.5
MT 4 × 3.25	3.25	MT 100 × 4.85	4.7
MT 4 × 3.1	3.1	MT 100 × 4.6	4.5
MT 3 × 2.2	2.2	MT 75 × 3.3	3.2
MT 3 × 1.85	1.85	MT 75 × 2.75	2.7
MT 2.5 × 9.45	9.45	MT 65 × 14.05	13.8
MT 2 × 3	3	MT 50 × 4.45	4.4

STRUCTURAL STEEL WEIGHT TABLE (IMPERIAL AND METRIC) *(cont.)*

ST (AMERICAN STANDARD TEES)

Imperial		Metric	
Designation	**Weight (lb/ft)**	**Designation**	**Weight (kg/m)**
ST 12 × 60.5	60.5	ST 305 × 90	88.3
ST 12 × 53	53	ST 305 × 79	77.3
ST 12 × 50	50	ST 305 × 74.5	73.0
ST 12 × 45	45	ST 305 × 67	65.7
ST 12 × 40	40	ST 305 × 59.5	58.4
ST 10 × 48	48	ST 255 × 71.5	70.0
ST 10 × 43	43	ST 255 × 64	62.8
ST 10 × 37.5	37.5	ST 255 × 56	54.7
ST 10 × 33	33	ST 255 × 49.1	48.2
ST 9 × 35	35	ST 230 × 52	51.1
ST 9 × 27.35	27.35	ST 230 × 40.7	39.9
ST 7.5 × 25	25	ST 190 × 37	36.5
ST 7.5 × 21.45	21.45	ST 190 × 32	31.3
ST 6 × 25	25	ST 155 × 37	36.5
ST 6 × 20.4	20.4	ST 155 × 30.35	29.8
ST 6 × 17.5	17.5	ST 155 × 26	25.5
ST 6 × 15.9	15.9	ST 155 × 23.65	23.2
ST 5 × 17.5	17.5	ST 125 × 26	25.5
ST 5 × 12.7	12.7	ST 125 × 18.9	18.5
ST 4 × 11.5	11.5	ST 100 × 17	16.8
ST 4 × 9.2	9.2	ST 100 × 13.7	13.4
ST 3 × 8.63	8.63	ST 75 × 12.85	12.6
ST 3 × 6.25	6.25	ST 75 × 9.3	9.1
ST 2.5 × 5	5	ST 65 × 7.5	7.3
ST 2 × 4.75	4.75	ST 50 × 7.05	6.9
ST 2 × 3.85	3.85	ST 50 × 5.75	5.6
ST 1.5 × 3.75	3.75	ST 37.5 × 5.6	5.5
ST 1.5 × 2.85	2.85	ST 37.5 × 4.25	4.2

STRUCTURAL STEEL WEIGHT TABLE (IMPERIAL AND METRIC) *(cont.)*

L (ANGLE WITH EQUAL LEGS)

Imperial		Metric	
Designation	**Weight (lb/ft)**	**Designation**	**Weight (kg/m)**
L 8 × 8 × 1⅛	57.2	L 203 × 203 × 28.6	83.5
L 8 × 8 × 1	51.3	L 203 × 203 × 25.4	74.9
L 8 × 8 × ⅞	45.3	L 203 × 203 × 22.2	66.1
L 8 × 8 × ¾	39.2	L 203 × 203 × 19	57.2
L 8 × 8 × ⅝	33	L 203 × 203 × 15.9	48.1
L 8 × 8 × 9⁄16	29.8	L 203 × 203 × 14.3	43.5
L 8 × 8 × ½	26.7	L 203 × 203 × 12.7	38.9
L 6 × 6 × 1	37.5	L 152 × 152 × 25.4	54.7
L 6 × 6 × ⅞	33.2	L 152 × 152 × 22.2	48.4
L 6 × 6 × ¾	28.8	L 152 × 152 × 19	42.0
L 6 × 6 × ⅝	24.3	L 152 × 152 × 15.9	35.4
L 6 × 6 × 9⁄16	22	L 152 × 152 × 14.3	32.0
L 6 × 6 × ½	19.6	L 152 × 152 × 12.7	28.7
L 6 × 6 × 7⁄16	17.3	L 152 × 152 × 11.1	25.2
L 6 × 6 × ⅜	14.9	L 152 × 152 × 9.5	21.7
L 6 × 6 × 5⁄16	12.5	L 152 × 152 × 7.9	18.2

STRUCTURAL STEEL WEIGHT TABLE
(IMPERIAL AND METRIC) *(cont.)*

L (ANGLE WITH EQUAL LEGS)

Imperial		Metric	
Designation	**Weight (lb/ft)**	**Designation**	**Weight (kg/m)**
L 5 × 5 × 7⁄8	27.3	L 127 × 127 × 22.2	39.8
L 5 × 5 × 3⁄4	23.7	L 127 × 127 × 19	34.6
L 5 × 5 × 5⁄8	20.1	L 127 × 127 × 15.9	29.3
L 5 × 5 × 1⁄2	16.3	L 127 × 127 × 12.7	23.8
L 5 × 5 × 7⁄16	14.4	L 127 × 127 × 11.1	21.0
L 5 × 5 × 3⁄8	12.4	L 127 × 127 × 9.5	18.1
L 5 × 5 × 5⁄16	10.4	L 127 × 127 × 7.9	15.2
L 4 × 4 × 3⁄4	18.5	L 102 × 102 × 19	27.0
L 4 × 4 × 5⁄8	15.7	L 102 × 102 × 15.9	22.9
L 4 × 4 × 1⁄2	12.7	L 102 × 102 × 12.7	18.6
L 4 × 4 × 7⁄16	11.2	L 102 × 102 × 11.1	16.4
L 4 × 4 × 3⁄8	9.72	L 102 × 102 × 9.5	14.2
L 4 × 4 × 5⁄16	8.16	L 102 × 102 × 7.9	11.9
L 4 × 4 × 1⁄4	6.58	L 102 × 102 × 6.4	9.6
L 3½ × 3½ × 1⁄2	11.1	L 89 × 89 × 12.7	16.2
L 3½ × 3½ × 7⁄16	9.82	L 89 × 89 × 11.1	14.3

STRUCTURAL STEEL WEIGHT TABLE (IMPERIAL AND METRIC) *(cont.)*

L (ANGLE WITH EQUAL LEGS)

Imperial		Metric	
Designation	Weight (lb/ft)	Designation	Weight (kg/m)
L 3½ × 3½ × ⅜	8.51	L 89 × 89 × 9.5	12.4
L 3½ × 3½ × ⁵⁄₁₆	7.16	L 89 × 89 × 7.9	10.5
L 3½ × 3½ × ¼	5.79	L 89 × 89 × 6.4	8.5
L 3 × 3 × ½	9.35	L 76 × 76 × 12.7	13.6
L 3 × 3 × ⁷⁄₁₆	8.28	L 76 × 76 × 11.1	12.1
L 3 × 3 × ⅜	7.17	L 76 × 76 × 9.5	10.5
L 3 × 3 × ⁵⁄₁₆	6.04	L 76 × 76 × 7.9	8.8
L 3 × 3 × ¼	4.89	L 76 × 76 × 6.4	7.1
L 3 × 3 × ³⁄₁₆	3.7	L 76 × 76 × 4.8	5.4
L 2½ × 2½ × ½	7.65	L 64 × 64 × 12.7	11.2
L 2½ × 2½ × ⅜	5.9	L 64 × 64 × 9.5	8.6
L 2½ × 2½ × ⁵⁄₁₆	4.98	L 64 × 64 × 7.9	7.3
L 2½ × 2½ × ¼	4.04	L 64 × 64 × 6.4	5.9
L 2½ × 2½ × ³⁄₁₆	3.06	L 64 × 64 × 4.8	4.5
L 2 × 2 × ⅜	4.65	L 51 × 51 × 9.5	6.8
L 2 × 2 × ⁵⁄₁₆	3.94	L 51 × 51 × 7.9	5.8
L 2 × 2 × ¼	3.21	L 51 × 51 × 6.4	4.7
L 2 × 2 × ³⁄₁₆	2.46	L 51 × 51 × 4.8	3.6
L 2 × 2 × ⅛	1.67	L 51 × 51 × 3.2	2.4

STRUCTURAL STEEL WEIGHT TABLE
(IMPERIAL AND METRIC) *(cont.)*

L (ANGLE WITH UNEQUAL LEGS)

Imperial		Metric	
Designation	**Weight (lb/ft)**	**Designation**	**Weight (kg/m)**
L 8 × 6 × 1	44.4	L 203 × 152 × 25.4	64.8
L 8 × 6 × 7/8	39.3	L 203 × 152 × 22.2	57.3
L 8 × 6 × 3/4	34	L 203 × 152 × 19	49.6
L 8 × 6 × 5/8	28.6	L 203 × 152 × 15.9	41.8
L 8 × 6 × 9/16	25.9	L 203 × 152 × 14.3	37.8
L 8 × 6 × 1/2	23.2	L 203 × 152 × 12.7	33.8
L 8 × 6 × 7/16	20.4	L 203 × 152 × 11.1	29.7
L 8 × 4 × 1	37.6	L 203 × 102 × 25.4	54.9
L 8 × 4 × 7/8	33.3	L 203 × 102 × 22.2	48.6
L 8 × 4 × 3/4	28.9	L 203 × 102 × 19	42.2
L 8 × 4 × 5/8	24.4	L 203 × 102 × 15.9	35.6
L 8 × 4 × 9/16	22.1	L 203 × 102 × 14.3	32.2
L 8 × 4 × 1/2	19.7	L 203 × 102 × 12.7	28.8
L 8 × 4 × 7/16	17.4	L 203 × 102 × 11.1	25.4
L 7 × 4 × 3/4	26.2	L 178 × 102 × 19	38.2
L 7 × 4 × 5/8	22.1	L 178 × 102 × 15.9	32.3
L 7 × 4 × 1/2	17.9	L 178 × 102 × 12.7	26.1
L 7 × 4 × 7/16	15.8	L 178 × 102 × 11.1	23.0
L 7 × 4 × 3/8	13.6	L 178 × 102 × 9.5	19.9
L 6 × 4 × 7/8	27.2	L 152 × 102 × 22.2	39.6
L 6 × 4 × 3/4	23.6	L 152 × 102 × 19	34.4
L 6 × 4 × 5/8	19.9	L 152 × 102 × 15.9	29.1
L 6 × 4 × 9/16	18.1	L 152 × 102 × 14.3	26.4
L 6 × 4 × 1/2	16.2	L 152 × 102 × 12.7	23.6

STRUCTURAL STEEL WEIGHT TABLE
(IMPERIAL AND METRIC) *(cont.)*

L (ANGLE WITH UNEQUAL LEGS)

Imperial		Metric	
Designation	**Weight (lb/ft)**	**Designation**	**Weight (kg/m)**
L 6 × 4 × 7/16	14.2	L 152 × 102 × 11.1	20.8
L 6 × 4 × 3/8	12.3	L 152 × 102 × 9.5	17.9
L 6 × 4 × 5/16	10.3	L 152 × 102 × 7.9	15.0
L 6 × 3½ × ½	15.4	L 152 × 89 × 12.7	22.4
L 6 × 3½ × 3/8	11.7	L 152 × 89 × 9.5	17.1
L 6 × 3½ × 5/16	9.83	L 152 × 89 × 7.9	14.3
L 5 × 3½ × 3/4	19.8	L 127 × 89 × 19	28.9
L 5 × 3½ × 5/8	16.8	L 127 × 89 × 15.9	24.5
L 5 × 3½ × ½	13.6	L 127 × 89 × 12.7	19.9
L 5 × 3½ × 3/8	10.4	L 127 × 89 × 9.5	15.1
L 5 × 3½ × 5/16	8.72	L 127 × 89 × 7.9	12.7
L 5 × 3½ × ¼	7.03	L 127 × 89 × 6.4	10.3
L 5 × 3 × ½	12.8	L 127 × 76 × 12.7	18.6
L 5 × 3 × 7/16	11.3	L 127 × 76 × 11.1	16.4
L 5 × 3 × 3/8	9.74	L 127 × 76 × 9.5	14.2
L 5 × 3 × 5/16	8.19	L 127 × 76 × 7.9	11.9
L 5 × 3 × ¼	6.6	L 127 × 76 × 6.4	9.6
L 4 × 3½ × ½	11.9	L 102 × 89 × 12.7	17.4
L 4 × 3½ × 3/8	9.1	L 102 × 89 × 9.5	13.3
L 4 × 3½ × 5/16	7.65	L 102 × 89 × 7.9	11.2
L 4 × 3½ × ¼	6.18	L 102 × 89 × 6.4	9.0
L 4 × 3 × 5/8	13.6	L 102 × 76 × 15.9	19.8
L 4 × 3 × ½	11.1	L 102 × 76 × 12.7	16.2
L 4 × 3 × 3/8	8.47	L 102 × 76 × 9.5	12.4

STRUCTURAL STEEL WEIGHT TABLE (IMPERIAL AND METRIC) *(cont.)*

L (ANGLE WITH UNEQUAL LEGS)

Imperial		Metric	
Designation	**Weight (lb/ft)**	**Designation**	**Weight (kg/m)**
L 4 × 3 × 5/16	7.12	L 102 × 76 × 7.9	10.4
L 4 × 3 × 1/4	5.75	L 102 × 76 × 6.4	8.4
L 3½ × 3 × 1/2	10.3	L 89 × 76 × 12.7	15.0
L 3½ × 3 × 7/16	9.09	L 89 × 76 × 11.1	13.3
L 3½ × 3 × 3/8	7.88	L 89 × 76 × 9.5	11.5
L 3½ × 3 × 5/16	6.65	L 89 × 76 × 7.9	9.7
L 3½ × 3 × 1/4	5.38	L 89 × 76 × 6.4	7.9
L 3½ × 2½ × 1/2	9.41	L 89 × 64 × 12.7	13.7
L 3½ × 2½ × 3/8	7.23	L 89 × 64 × 9.5	10.5
L 3½ × 2½ × 5/16	6.1	L 89 × 64 × 7.9	8.9
L 3½ × 2½ × 1/4	4.94	L 89 × 64 × 6.4	7.2
L 3 × 2½ × 1/2	8.53	L 76 × 64 × 12.7	12.4
L 3 × 2½ × 7/16	7.56	L 76 × 64 × 11.1	11.0
L 3 × 2½ × 3/8	6.56	L 76 × 64 × 9.5	9.6
L 3 × 2½ × 5/16	5.54	L 76 × 64 × 7.9	8.1
L 3 × 2½ × 1/4	4.49	L 76 × 64 × 6.4	6.6
L 3 × 2½ × 3/16	3.41	L 76 × 64 × 4.8	5.0
L 3 × 2 × 1/2	7.7	L 76 × 51 × 12.7	11.2
L 3 × 2 × 3/8	5.95	L 76 × 51 × 9.5	8.7
L 3 × 2 × 5/16	5.03	L 76 × 51 × 7.9	7.3
L 3 × 2 × 1/4	4.09	L 76 × 51 × 6.4	6.0
L 3 × 2 × 3/16	3.12	L 76 × 51 × 4.8	4.6
L 2½ × 2 × 3/8	5.3	L 64 × 51 × 9.5	7.7
L 2½ × 2 × 5/16	4.49	L 64 × 51 × 7.9	6.6
L 2½ × 2 × 1/4	3.65	L 64 × 51 × 6.4	5.3
L 2½ × 2 × 3/16	2.78	L 64 × 51 × 4.8	4.1

STRUCTURAL STEEL WEIGHT TABLE (IMPERIAL AND METRIC) (cont.)

2L (DOUBLE ANGLES WITH EQUAL LEGS)

Imperial		Metric	
Designation	Weight (lb/ft)	Designation	Weight (kg/m)
2L 8 × 8 × 1⅛	114	2L 203 × 203 × 28.6	167.0
2L 8 × 8 × 1⅛ × ⅜	114	2L 203 × 203 × 28.6 × 9	167.0
2L 8 × 8 × 1⅛ × ¾	114	2L 203 × 203 × 28.6 × 19	167.0
2L 8 × 8 × 1	103	2L 203 × 203 × 25.4	150.0
2L 8 × 8 × 1 × ⅜	103	2L 203 × 203 × 25.4 × 9	150.0
2L 8 × 8 × 1 × ¾	103	2L 203 × 203 × 25.4 × 19	150.0
2L 8 × 8 × ⅞	90.6	2L 203 × 203 × 22.2	132.0
2L 8 × 8 × ⅞ × ⅜	90.6	2L 203 × 203 × 22.2 × 9	132.0
2L 8 × 8 × ⅞ × ¾	90.6	2L 203 × 203 × 22.2 × 19	132.0
2L 8 × 8 × ¾	78.4	2L 203 × 203 × 19	114.0
2L 8 × 8 × ¾ × ⅜	78.4	2L 203 × 203 × 19 × 9	114.0
2L 8 × 8 × ¾ × ¾	78.4	2L 203 × 203 × 19 × 19	114.0
2L 8 × 8 × ⅝	66	2L 203 × 203 × 15.9	96.3
2L 8 × 8 × ⅝ × ⅜	66	2L 203 × 203 × 15.9 × 9	96.3
2L 8 × 8 × ⅝ × ¾	66	2L 203 × 203 × 15.9 × 19	96.3
2L 8 × 8 × ⁹⁄₁₆	59.7	2L 203 × 203 × 14.3	87.1
2L 8 × 8 × ⁹⁄₁₆ × ⅜	59.7	2L 203 × 203 × 14.3 × 9	87.1
2L 8 × 8 × ⁹⁄₁₆ × ¾	59.7	2L 203 × 203 × 14.3 × 19	87.1
2L 8 × 8 × ½	53.3	2L 203 × 203 × 12.7	77.8
2L 8 × 8 × ½ × ⅜	53.3	2L 203 × 203 × 12.7 × 9	77.8
2L 8 × 8 × ½ × ¾	53.3	2L 203 × 203 × 12.7 × 19	77.8
2L 6 × 6 × 1	75	2L 152 × 152 × 25.4	109.0
2L 6 × 6 × 1 × ⅜	75	2L 152 × 152 × 25.4 × 9	109.0
2L 6 × 6 × 1 × ¾	75	2L 152 × 152 × 25.4 × 19	109.0
2L 6 × 6 × ⅞	66.4	2L 152 × 152 × 22.2	96.9
2L 6 × 6 × ⅞ × ⅜	66.4	2L 152 × 152 × 22.2 × 9	96.9
2L 6 × 6 × ⅞ × ¾	66.4	2L 152 × 152 × 22.2 × 19	96.9
2L 6 × 6 × ¾	57.6	2L 152 × 152 × 19	84.0
2L 6 × 6 × ¾ × ⅜	57.6	2L 152 × 152 × 19 × 9	84.0
2L 6 × 6 × ¾ × ¾	57.6	2L 152 × 152 × 19 × 19	84.0
2L 6 × 6 × ⅝	48.5	2L 152 × 152 × 15.9	70.8
2L 6 × 6 × ⅝ × ⅜	48.5	2L 152 × 152 × 15.9 × 9	70.8

STRUCTURAL STEEL WEIGHT TABLE (IMPERIAL AND METRIC) *(cont.)*

2L (DOUBLE ANGLES WITH EQUAL LEGS)

Imperial		Metric	
Designation	**Weight (lb/ft)**	**Designation**	**Weight (kg/m)**
2L 6 × 6 × ⅝ × ¾	48.5	2L 152 × 152 × 15.9 × 19	70.8
2L 6 × 6 × 9/16	43.9	2L 152 × 152 × 14.3	64.1
2L 6 × 6 × 9/16 × ⅜	43.9	2L 152 × 152 × 14.3 × 9	64.1
2L 6 × 6 × 9/16 × ¾	43.9	2L 152 × 152 × 14.3 × 19	64.1
2L 6 × 6 × ½	39.3	2L 152 × 152 × 12.7	57.3
2L 6 × 6 × ½ × ⅜	39.3	2L 152 × 152 × 12.7 × 9	57.3
2L 6 × 6 × ½ × ¾	39.3	2L 152 × 152 × 12.7 × 19	57.3
2L 6 × 6 × 7/16	34.6	2L 152 × 152 × 11.1	50.4
2L 6 × 6 × 7/16 × ⅜	34.6	2L 152 × 152 × 11.1 × 9	50.4
2L 6 × 6 × 7/16 × ¾	34.6	2L 152 × 152 × 11.1 × 19	50.4
2L 6 × 6 × ⅜	29.8	2L 152 × 152 × 9.5	43.5
2L 6 × 6 × ⅜ × ⅜	29.8	2L 152 × 152 × 9.5 × 9	43.5
2L 6 × 6 × ⅜ × ¾	29.8	2L 152 × 152 × 9.5 × 19	43.5
2L 6 × 6 × 5/16	25	2L 152 × 152 × 7.9	36.5
2L 6 × 6 × 5/16 × ⅜	25	2L 152 × 152 × 7.9 × 9	36.5
2L 6 × 6 × 5/16 × ¾	25	2L 152 × 152 × 7.9 × 19	36.5
2L 5 × 5 × ⅞	54.6	2L 127 × 127 × 22.2	79.7
2L 5 × 5 × ⅞ × ⅜	54.6	2L 127 × 127 × 22.2 × 9	79.7
2L 5 × 5 × ⅞ × ¾	54.6	2L 127 × 127 × 22.2 × 19	79.7
2L 5 × 5 × ¾	47.5	2L 127 × 127 × 19	69.3
2L 5 × 5 × ¾ × ⅜	47.5	2L 127 × 127 × 19 × 9	69.3
2L 5 × 5 × ¾ × ¾	47.5	2L 127 × 127 × 19 × 19	69.3
2L 5 × 5 × ⅝	40.1	2L 127 × 127 × 15.9	58.6
2L 5 × 5 × ⅝ × ⅜	40.1	2L 127 × 127 × 15.9 × 9	58.6
2L 5 × 5 × ⅝ × ¾	40.1	2L 127 × 127 × 15.9 × 19	58.6
2L 5 × 5 × ½	32.6	2L 127 × 127 × 12.7	47.6
2L 5 × 5 × ½ × ⅜	32.6	2L 127 × 127 × 12.7 × 9	47.6
2L 5 × 5 × ½ × ¾	32.6	2L 127 × 127 × 12.7 × 19	47.6
2L 5 × 5 × 7/16	28.7	2L 127 × 127 × 11.1	41.9
2L 5 × 5 × 7/16 × ⅜	28.7	2L 127 × 127 × 11.1 × 9	41.9
2L 5 × 5 × 7/16 × ¾	28.7	2L 127 × 127 × 11.1 × 19	41.9
2L 5 × 5 × ⅜	24.8	2L 127 × 127 × 9.5	36.2

STRUCTURAL STEEL WEIGHT TABLE (IMPERIAL AND METRIC) *(cont.)*

2L (DOUBLE ANGLES WITH EQUAL LEGS)

Imperial		Metric	
Designation	**Weight (lb/ft)**	**Designation**	**Weight (kg/m)**
2L 5 × 5 × ⅜ × ⅜	24.8	2L 127 × 127 × 9.5 × 9	36.2
2L 5 × 5 × ⅜ × ¾	24.8	2L 127 × 127 × 9.5 × 19	36.2
2L 5 × 5 × 5⁄16	20.9	2L 127 × 127 × 7.9	30.4
2L 5 × 5 × 5⁄16 × ⅜	20.9	2L 127 × 127 × 7.9 × 9	30.4
2L 5 × 5 × 5⁄16 × ¾	20.9	2L 127 × 127 × 7.9 × 19	30.4
2L 4 × 4 × ¾	37	2L 102 × 102 × 19	54.0
2L 4 × 4 × ¾ × ⅜	37	2L 102 × 102 × 19 × 9	54.0
2L 4 × 4 × ¾ × ¾	37	2L 102 × 102 × 19 × 19	54.0
2L 4 × 4 × ⅝	31.3	2L 102 × 102 × 15.9	45.7
2L 4 × 4 × ⅝ × ⅜	31.3	2L 102 × 102 × 15.9 × 9	45.7
2L 4 × 4 × ⅝ × ¾	31.3	2L 102 × 102 × 15.9 × 19	45.7
2L 4 × 4 × ½	25.5	2L 102 × 102 × 12.7	37.2
2L 4 × 4 × ½ × ⅜	25.5	2L 102 × 102 × 12.7 × 9	37.2
2L 4 × 4 × ½ × ¾	25.5	2L 102 × 102 × 12.7 × 19	37.2
2L 4 × 4 × 7⁄16	22.5	2L 102 × 102 × 11.1	32.8
2L 4 × 4 × 7⁄16 × ⅜	22.5	2L 102 × 102 × 11.1 × 9	32.8
2L 4 × 4 × 7⁄16 × ¾	22.5	2L 102 × 102 × 11.1 × 19	32.8
2L 4 × 4 × ⅜	19.4	2L 102 × 102 × 9.5	28.4
2L 4 × 4 × ⅜ × ⅜	19.4	2L 102 × 102 × 9.5 × 9	28.4
2L 4 × 4 × ⅜ × ¾	19.4	2L 102 × 102 × 9.5 × 19	28.4
2L 4 × 4 × 5⁄16	16.3	2L 102 × 102 × 7.9	23.8
2L 4 × 4 × 5⁄16 × ⅜	16.3	2L 102 × 102 × 7.9 × 9	23.8
2L 4 × 4 × 5⁄16 × ¾	16.3	2L 102 × 102 × 7.9 × 19	23.8
2L 4 × 4 × ¼	13.2	2L 102 × 102 × 6.4	19.2
2L 4 × 4 × ¼ × ⅜	13.2	2L 102 × 102 × 6.4 × 9	19.2
2L 4 × 4 × ¼ × ¾	13.2	2L 102 × 102 × 6.4 × 19	19.2
2L 3½ × 3½ × ½	22.2	2L 89 × 89 × 12.7	32.4
2L 3½ × 3½ × ½ × ⅜	22.2	2L 89 × 89 × 12.7 × 9	32.4
2L 3½ × 3½ × ½ × ¾	22.2	2L 89 × 89 × 12.7 × 19	32.4
2L 3½ × 3½ × 7⁄16	19.6	2L 89 × 89 × 11.1	28.7
2L 3½ × 3½ × 7⁄16 × ⅜	19.6	2L 89 × 89 × 11.1 × 9	28.7
2L 3½ × 3½ × 7⁄16 × ¾	19.6	2L 89 × 89 × 11.1 × 19	28.7

STRUCTURAL STEEL WEIGHT TABLE (IMPERIAL AND METRIC) *(cont.)*

2L (DOUBLE ANGLES WITH EQUAL LEGS)

Imperial		Metric	
Designation	**Weight (lb/ft)**	**Designation**	**Weight (kg/m)**
2L $3\frac{1}{2} \times 3\frac{1}{2} \times \frac{3}{8}$	17	2L 89 × 89 × 9.5	24.8
2L $3\frac{1}{2} \times 3\frac{1}{2} \times \frac{3}{8} \times \frac{3}{8}$	17	2L 89 × 89 × 9.5 × 9	24.8
2L $3\frac{1}{2} \times 3\frac{1}{2} \times \frac{3}{8} \times \frac{3}{4}$	17	2L 89 × 89 × 9.5 × 19	24.8
2L $3\frac{1}{2} \times 3\frac{1}{2} \times \frac{5}{16}$	14.3	2L 89 × 89 × 7.9	20.9
2L $3\frac{1}{2} \times 3\frac{1}{2} \times \frac{5}{16} \times \frac{3}{8}$	14.3	2L 89 × 89 × 7.9 × 9	20.9
2L $3\frac{1}{2} \times 3\frac{1}{2} \times \frac{5}{16} \times \frac{3}{4}$	14.3	2L 89 × 89 × 7.9 × 19	20.9
2L $3\frac{1}{2} \times 3\frac{1}{2} \times \frac{1}{4}$	11.6	2L 89 × 89 × 6.4	16.9
2L $3\frac{1}{2} \times 3\frac{1}{2} \times \frac{1}{4} \times \frac{3}{8}$	11.6	2L 89 × 89 × 6.4 × 9	16.9
2L $3\frac{1}{2} \times 3\frac{1}{2} \times \frac{1}{4} \times \frac{3}{4}$	11.6	2L 89 × 89 × 6.4 × 19	16.9
2L $3 \times 3 \times \frac{1}{2}$	18.7	2L 76 × 76 × 12.7	27.3
2L $3 \times 3 \times \frac{1}{2} \times \frac{3}{8}$	18.7	2L 76 × 76 × 12.7 × 9	27.3
2L $3 \times 3 \times \frac{1}{2} \times \frac{3}{4}$	18.7	2L 76 × 76 × 12.7 × 19	27.3
2L $3 \times 3 \times \frac{7}{16}$	16.6	2L 76 × 76 × 11.1	24.2
2L $3 \times 3 \times \frac{7}{16} \times \frac{3}{8}$	16.6	2L 76 × 76 × 11.1 × 9	24.2
2L $3 \times 3 \times \frac{7}{16} \times \frac{3}{4}$	16.6	2L 76 × 76 × 11.1 × 19	24.2
2L $3 \times 3 \times \frac{3}{8}$	14.3	2L 76 × 76 × 9.5	20.9
2L $3 \times 3 \times \frac{3}{8} \times \frac{3}{8}$	14.3	2L 76 × 76 × 9.5 × 9	20.9
2L $3 \times 3 \times \frac{3}{8} \times \frac{3}{4}$	14.3	2L 76 × 76 × 9.5 × 19	20.9
2L $3 \times 3 \times \frac{5}{16}$	12.1	2L 76 × 76 × 7.9	17.6
2L $3 \times 3 \times \frac{5}{16} \times \frac{3}{8}$	12.1	2L 76 × 76 × 7.9 × 9	17.6
2L $3 \times 3 \times \frac{5}{16} \times \frac{3}{4}$	12.1	2L 76 × 76 × 7.9 × 19	17.6
2L $3 \times 3 \times \frac{1}{4}$	9.77	2L 76 × 76 × 6.4	14.3
2L $3 \times 3 \times \frac{1}{4} \times \frac{3}{8}$	9.77	2L 76 × 76 × 6.4 × 9	14.3
2L $3 \times 3 \times \frac{1}{4} \times \frac{3}{4}$	9.77	2L 76 × 76 × 6.4 × 19	14.3
2L $3 \times 3 \times \frac{3}{16}$	7.41	2L 76 × 76 × 4.8	10.8
2L $3 \times 3 \times \frac{3}{16} \times \frac{3}{8}$	7.41	2L 76 × 76 × 4.8 × 9	10.8
2L $3 \times 3 \times \frac{3}{16} \times \frac{3}{4}$	7.41	2L 76 × 76 × 4.8 × 19	10.8
2L $2\frac{1}{2} \times 2\frac{1}{2} \times \frac{1}{2}$	15.3	2L 64 × 64 × 12.7	22.3
2L $2\frac{1}{2} \times 2\frac{1}{2} \times \frac{1}{2} \times \frac{3}{8}$	15.3	2L 64 × 64 × 12.7 × 9	22.3
2L $2\frac{1}{2} \times 2\frac{1}{2} \times \frac{1}{2} \times \frac{3}{4}$	15.3	2L 64 × 64 × 12.7 × 19	22.3
2L $2\frac{1}{2} \times 2\frac{1}{2} \times \frac{3}{8}$	11.8	2L 64 × 64 × 9.5	17.2
2L $2\frac{1}{2} \times 2\frac{1}{2} \times \frac{3}{8} \times \frac{3}{8}$	11.8	2L 64 × 64 × 9.5 × 9	17.2

STRUCTURAL STEEL WEIGHT TABLE (IMPERIAL AND METRIC) (cont.)

2L (DOUBLE ANGLES WITH EQUAL LEGS)

Imperial		Metric	
Designation	Weight (lb/ft)	Designation	Weight (kg/m)
2L $2\frac{1}{2} \times 2\frac{1}{2} \times \frac{3}{8} \times \frac{3}{4}$	11.8	2L $64 \times 64 \times 9.5 \times 19$	17.2
2L $2\frac{1}{2} \times 2\frac{1}{2} \times \frac{5}{16}$	9.96	2L $64 \times 64 \times 7.9$	14.5
2L $2\frac{1}{2} \times 2\frac{1}{2} \times \frac{5}{16} \times \frac{3}{8}$	9.96	2L $64 \times 64 \times 7.9 \times 9$	14.5
2L $2\frac{1}{2} \times 2\frac{1}{2} \times \frac{5}{16} \times \frac{3}{4}$	9.96	2L $64 \times 64 \times 7.9 \times 19$	14.5
2L $2\frac{1}{2} \times 2\frac{1}{2} \times \frac{1}{4}$	8.07	2L $64 \times 64 \times 6.4$	11.8
2L $2\frac{1}{2} \times 2\frac{1}{2} \times \frac{1}{4} \times \frac{3}{8}$	8.07	2L $64 \times 64 \times 6.4 \times 9$	11.8
2L $2\frac{1}{2} \times 2\frac{1}{2} \times \frac{1}{4} \times \frac{3}{4}$	8.07	2L $64 \times 64 \times 6.4 \times 19$	11.8
2L $2\frac{1}{2} \times 2\frac{1}{2} \times \frac{3}{16}$	6.13	2L $64 \times 64 \times 4.8$	8.9
2L $2\frac{1}{2} \times 2\frac{1}{2} \times \frac{3}{16} \times \frac{3}{8}$	6.13	2L $64 \times 64 \times 4.8 \times 9$	8.9
2L $2\frac{1}{2} \times 2\frac{1}{2} \times \frac{3}{16} \times \frac{3}{4}$	6.13	2L $64 \times 64 \times 4.8 \times 19$	8.9
2L $2 \times 2 \times \frac{3}{8}$	9.3	2L $51 \times 51 \times 9.5$	13.6
2L $2 \times 2 \times \frac{3}{8} \times \frac{3}{8}$	9.3	2L $51 \times 51 \times 9.5 \times 9$	13.6
2L $2 \times 2 \times \frac{3}{8} \times \frac{3}{4}$	9.3	2L $51 \times 51 \times 9.5 \times 19$	13.6
2L $2 \times 2 \times \frac{5}{16}$	7.89	2L $51 \times 51 \times 7.9$	11.5
2L $2 \times 2 \times \frac{5}{16} \times \frac{3}{8}$	7.89	2L $51 \times 51 \times 7.9 \times 9$	11.5
2L $2 \times 2 \times \frac{5}{16} \times \frac{3}{4}$	7.89	2L $51 \times 51 \times 7.9 \times 19$	11.5
2L $2 \times 2 \times \frac{1}{4}$	6.43	2L $51 \times 51 \times 6.4$	9.4
2L $2 \times 2 \times \frac{1}{4} \times \frac{3}{8}$	6.43	2L $51 \times 51 \times 6.4 \times 9$	9.4
2L $2 \times 2 \times \frac{1}{4} \times \frac{3}{4}$	6.43	2L $51 \times 51 \times 6.4 \times 19$	9.4
2L $2 \times 2 \times \frac{3}{16}$	4.91	2L $51 \times 51 \times 4.8$	7.2
2L $2 \times 2 \times \frac{3}{16} \times \frac{3}{8}$	4.91	2L $51 \times 51 \times 4.8 \times 9$	7.2
2L $2 \times 2 \times \frac{3}{16} \times \frac{3}{4}$	4.91	2L $51 \times 51 \times 4.8 \times 19$	7.2
2L $2 \times 2 \times \frac{1}{8}$	3.34	2L $51 \times 51 \times 3.2$	4.9
2L $2 \times 2 \times \frac{1}{8} \times \frac{3}{8}$	3.34	2L $51 \times 51 \times 3.2 \times 9$	4.9
2L $2 \times 2 \times \frac{1}{8} \times \frac{3}{4}$	3.34	2L $51 \times 51 \times 3.2 \times 19$	4.9

STRUCTURAL STEEL WEIGHT TABLE (IMPERIAL AND METRIC) *(cont.)*

2L-LLBB (DOUBLE ANGLES WITH LONG LEGS BACK-TO-BACK)

Imperial		Metric	
Designation	**Weight (lb/ft)**	**Designation**	**Weight (kg/m)**
2L 8 × 6 × 1 LLBB	88.8	2L 203 × 152 × 25.4 LLBB	130.0
2L 8 × 6 × 1 × 3/8 LLBB	88.8	2L 203 × 152 × 25.4 × 9 LLBB	130.0
2L 8 × 6 × 1 × 3/4 LLBB	88.8	2L 203 × 152 × 25.4 × 19 LLBB	130.0
2L 8 × 6 × 7/8 LLBB	78.5	2L 203 × 152 × 22.2 LLBB	115.0
2L 8 × 6 × 7/8 × 3/8 LLBB	78.5	2L 203 × 152 × 22.2 × 9 LLBB	115.0
2L 8 × 6 × 7/8 × 3/4 LLBB	78.5	2L 203 × 152 × 22.2 × 19 LLBB	115.0
2L 8 × 6 × 3/4 LLBB	68	2L 203 × 152 × 19 LLBB	99.2
2L 8 × 6 × 3/4 × 3/8 LLBB	68	2L 203 × 152 × 19 × 9 LLBB	99.2
2L 8 × 6 × 3/4 × 3/4 LLBB	68	2L 203 × 152 × 19 × 19 LLBB	99.2
2L 8 × 6 × 5/8 LLBB	57.3	2L 203 × 152 × 15.9 LLBB	83.6
2L 8 × 6 × 5/8 × 3/8 LLBB	57.3	2L 203 × 152 × 15.9 × 9 LLBB	83.6
2L 8 × 6 × 5/8 × 3/4 LLBB	57.3	2L 203 × 152 × 15.9 × 19 LLBB	83.6
2L 8 × 6 × 9/16 LLBB	51.8	2L 203 × 152 × 14.3 LLBB	75.6
2L 8 × 6 × 9/16 × 3/8 LLBB	51.8	2L 203 × 152 × 14.3 × 9 LLBB	75.6
2L 8 × 6 × 9/16 × 3/4 LLBB	51.8	2L 203 × 152 × 14.3 × 19 LLBB	75.6
2L 8 × 6 × 1/2 LLBB	46.3	2L 203 × 152 × 12.7 LLBB	67.6
2L 8 × 6 × 1/2 × 3/8 LLBB	46.3	2L 203 × 152 × 12.7 × 9 LLBB	67.6
2L 8 × 6 × 1/2 × 3/4 LLBB	46.3	2L 203 × 152 × 12.7 × 19 LLBB	67.6
2L 8 × 6 × 7/16 LLBB	40.7	2L 203 × 152 × 11.1 LLBB	59.5
2L 8 × 6 × 7/16 × 3/8 LLBB	40.7	2L 203 × 152 × 11.1 × 9 LLBB	59.5
2L 8 × 6 × 7/16 × 3/4 LLBB	40.7	2L 203 × 152 × 11.1 × 19 LLBB	59.5
2L 8 × 4 × 1 LLBB	75.2	2L 203 × 102 × 25.4 LLBB	110.0

STRUCTURAL STEEL WEIGHT TABLE (IMPERIAL AND METRIC) *(cont.)*

2L-LLBB (DOUBLE ANGLES WITH LONG LEGS BACK-TO-BACK)

Imperial		Metric	
Designation	**Weight (lb/ft)**	**Designation**	**Weight (kg/m)**
2L 8 × 4 × 1 × ⅜ LLBB	75.2	2L 203 × 102 × 25.4 × 9 LLBB	110.0
2L 8 × 4 × 1 × ¾ LLBB	75.2	2L 203 × 102 × 25.4 × 19 LLBB	110.0
2L 8 × 4 × ⅞ LLBB	66.6	2L 203 × 102 × 22.2 LLBB	97.2
2L 8 × 4 × ⅞ × ⅜ LLBB	66.6	2L 203 × 102 × 22.2 × 9 LLBB	97.2
2L 8 × 4 × ⅞ × ¾ LLBB	66.6	2L 203 × 102 × 22.2 × 19 LLBB	97.2
2L 8 × 4 × ¾ LLBB	57.8	2L 203 × 102 × 19 LLBB	84.3
2L 8 × 4 × ¾ × ⅜ LLBB	57.8	2L 203 × 102 × 19 × 9 LLBB	84.3
2L 8 × 4 × ¾ × ¾ LLBB	57.8	2L 203 × 102 × 19 × 19 LLBB	84.3
2L 8 × 4 × ⅝ LLBB	48.7	2L 203 × 102 × 15.9 LLBB	71.1
2L 8 × 4 × ⅝ × ⅜ LLBB	48.7	2L 203 × 102 × 15.9 × 9 LLBB	71.1
2L 8 × 4 × ⅝ × ¾ LLBB	48.7	2L 203 × 102 × 15.9 × 19 LLBB	71.1
2L 8 × 4 × 9⁄16 LLBB	44.1	2L 203 × 102 × 14.3 LLBB	64.4
2L 8 × 4 × 9⁄16 × 3/8 LLBB	44.1	2L 203 × 102 × 14.3 × 9 LLBB	64.4
2L 8 × 4 × 9⁄16 × ¾ LLBB	44.1	2L 203 × 102 × 14.3 × 19 LLBB	64.4
2L 8 × 4 × ½ LLBB	39.5	2L 203 × 102 × 12.7 LLBB	57.6
2L 8 × 4 × ½ × ⅜ LLBB	39.5	2L 203 × 102 × 12.7 × 9 LLBB	57.6
2L 8 × 4 × ½ × ¾ LLBB	39.5	2L 203 × 102 × 12.7 × 19 LLBB	57.6
2L 8 × 4 × 7⁄16 LLBB	34.8	2L 203 × 102 × 11.1 LLBB	50.8
2L 8 × 4 × 7⁄16 × ⅜ LLBB	34.8	2L 203 × 102 × 11.1 × 9 LLBB	50.8
2L 8 × 4 × 7⁄16 × ¾ LLBB	34.8	2L 203 × 102 × 11.1 × 19 LLBB	50.8
2L 7 × 4 × ¾ LLBB	52.4	2L 178 × 102 × 19 LLBB	76.5
2L 7 × 4 × ¾ × ⅜ LLBB	52.4	2L 178 × 102 × 19 × 9 LLBB	76.5

STRUCTURAL STEEL WEIGHT TABLE (IMPERIAL AND METRIC) *(cont.)*

2L-LLBB (DOUBLE ANGLES WITH LONG LEGS BACK-TO-BACK)

Imperial		Metric	
Designation	**Weight (lb/ft)**	**Designation**	**Weight (kg/m)**
2L 7 × 4 × ¾ × ¾ LLBB	52.4	2L 178 × 102 × 19 × 19 LLBB	76.5
2L 7 × 4 × ⅝ LLBB	44.2	2L 178 × 102 × 15.9 LLBB	64.5
2L 7 × 4 × ⅝ × ⅜ LLBB	44.2	2L 178 × 102 × 15.9 × 9 LLBB	64.5
2L 7 × 4 × ⅝ × ¾ LLBB	44.2	2L 178 × 102 × 15.9 × 19 LLBB	64.5
2L 7 × 4 × ½ LLBB	35.8	2L 178 × 102 × 12.7 LLBB	52.3
2L 7 × 4 × ½ × ⅜ LLBB	35.8	2L 178 × 102 × 12.7 × 9 LLBB	52.3
2L 7 × 4 × ½ × ¾ LLBB	35.8	2L 178 × 102 × 12.7 × 19 LLBB	52.3
2L 7 × 4 × ⁷⁄₁₆ LLBB	31.5	2L 178 × 102 × 11.1 LLBB	46.0
2L 7 × 4 × ⁷⁄₁₆ × ⅜ LLBB	31.5	2L 178 × 102 × 11.1 × 9 LLBB	46.0
2L 7 × 4 × ⁷⁄₁₆ × ¾ LLBB	31.5	2L 178 × 102 × 11.1 × 19 LLBB	46.0
2L 7 × 4 × ⅜ LLBB	27.2	2L 178 × 102 × 9.5 LLBB	39.7
2L 7 × 4 × ⅜ × ⅜ LLBB	27.2	2L 178 × 102 × 9.5 × 9 LLBB	39.7
2L 7 × 4 × ⅜ × ¾ LLBB	27.2	2L 178 × 102 × 9.5 × 19 LLBB	39.7
2L 6 × 4 × ⅞ LLBB	54.3	2L 152 × 102 × 22.2 LLBB	79.3
2L 6 × 4 × ⅞ × ⅜ LLBB	54.3	2L 152 × 102 × 22.2 × 9 LLBB	79.3
2L 6 × 4 × ⅞ × ¾ LLBB	54.3	2L 152 × 102 × 22.2 × 19 LLBB	79.3
2L 6 × 4 × ¾ LLBB	47.2	2L 152 × 102 × 19 LLBB	68.9
2L 6 × 4 × ¾ × ⅜ LLBB	47.2	2L 152 × 102 × 19 × 9 LLBB	68.9
2L 6 × 4 × ¾ × ¾ LLBB	47.2	2L 152 × 102 × 19 × 19 LLBB	68.9
2L 6 × 4 × ⅝ LLBB	39.9	2L 152 × 102 × 15.9 LLBB	58.2
2L 6 × 4 × ⅝ × ⅜ LLBB	39.9	2L 152 × 102 × 15.9 × 9 LLBB	58.2
2L 6 × 4 × ⅝ × ¾ LLBB	39.9	2L 152 × 102 × 15.9 × 19 LLBB	58.2

STRUCTURAL STEEL WEIGHT TABLE (IMPERIAL AND METRIC) *(cont.)*

2L-LLBB (DOUBLE ANGLES WITH LONG LEGS BACK-TO-BACK)

Imperial		Metric	
Designation	**Weight (lb/ft)**	**Designation**	**Weight (kg/m)**
2L 6 × 4 × ⁹⁄₁₆ LLBB	36.1	2L 152 × 102 × 14.3 LLBB	52.7
2L 6 × 4 × ⁹⁄₁₆ × ³⁄₈ LLBB	36.1	2L 152 × 102 × 14.3 × 9 LLBB	52.7
2L 6 × 4 × ⁹⁄₁₆ × ³⁄₄ LLBB	36.1	2L 152 × 102 × 14.3 × 19 LLBB	52.7
2L 6 × 4 × ½ LLBB	32.3	2L 152 × 102 × 12.7 LLBB	47.2
2L 6 × 4 × ½ × ³⁄₈ LLBB	32.3	2L 152 × 102 × 12.7 × 9 LLBB	47.2
2L 6 × 4 × ½ × ³⁄₄ LLBB	32.3	2L 152 × 102 × 12.7 × 19 LLBB	47.2
2L 6 × 4 × ⁷⁄₁₆ LLBB	28.5	2L 152 × 102 × 11.1 LLBB	41.5
2L 6 × 4 × ⁷⁄₁₆ × ³⁄₈ LLBB	28.5	2L 152 × 102 × 11.1 × 9 LLBB	41.5
2L 6 × 4 × ⁷⁄₁₆ × ³⁄₄ LLBB	28.5	2L 152 × 102 × 11.1 × 19 LLBB	41.5
2L 6 × 4 × ³⁄₈ LLBB	24.6	2L 152 × 102 × 9.5 LLBB	35.8
2L 6 × 4 × ³⁄₈ × ³⁄₈ LLBB	24.6	2L 152 × 102 × 9.5 × 9 LLBB	35.8
2L 6 × 4 × ³⁄₈ × ³⁄₄ LLBB	24.6	2L 152 × 102 × 9.5 × 19 LLBB	35.8
2L 6 × 4 × ⁵⁄₁₆ LLBB	20.6	2L 152 × 102 × 7.9 LLBB	30.1
2L 6 × 4 × ⁵⁄₁₆ × ³⁄₈ LLBB	20.6	2L 152 × 102 × 7.9 × 9 LLBB	30.1
2L 6 × 4 × ⁵⁄₁₆ × ³⁄₄ LLBB	20.6	2L 152 × 102 × 7.9 × 19 LLBB	30.1
2L 6 × 3½ × ½ LLBB	30.7	2L 152 × 89 × 12.7 LLBB	44.9
2L 6 × 3½ × ½ × ³⁄₈ LLBB	30.7	2L 152 × 89 × 12.7 × 9 LLBB	44.9
2L 6 × 3½ × ½ × ³⁄₄ LLBB	30.7	2L 152 × 89 × 12.7 × 19 LLBB	44.9
2L 6 × 3½ × ³⁄₈ LLBB	23.4	2L 152 × 89 × 9.5 LLBB	34.2
2L 6 × 3½ × ³⁄₈ × ³⁄₈ LLBB	23.4	2L 152 × 89 × 9.5 × 9 LLBB	34.2
2L 6 × 3½ × ³⁄₈ × ³⁄₄ LLBB	23.4	2L 152 × 89 × 9.5 × 19 LLBB	34.2
2L 6 × 3½ × ⁵⁄₁₆ LLBB	19.7	2L 152 × 89 × 7.9 LLBB	28.7

STRUCTURAL STEEL WEIGHT TABLE (IMPERIAL AND METRIC) *(cont.)*

2L-LLBB (DOUBLE ANGLES WITH LONG LEGS BACK-TO-BACK)

Imperial		Metric	
Designation	**Weight (lb/ft)**	**Designation**	**Weight (kg/m)**
2L 6 × 3½ × ⁵⁄₁₆ × ⅜ LLBB	19.7	2L 152 × 89 × 7.9 × 9 LLBB	28.7
2L 6 × 3½ × ⁵⁄₁₆ × ¾ LLBB	19.7	2L 152 × 89 × 7.9 × 19 LLBB	28.7
2L 5 × 3½ × ¾ LLBB	39.6	2L 127 × 89 × 19 LLBB	57.8
2L 5 × 3½ × ¾ × ⅜ LLBB	39.6	2L 127 × 89 × 19 × 9 LLBB	57.8
2L 5 × 3½ × ¾ × ¾ LLBB	39.6	2L 127 × 89 × 19 × 19 LLBB	57.8
2L 5 × 3½ × ⅝ LLBB	33.5	2L 127 × 89 × 15.9 LLBB	48.9
2L 5 × 3½ × ⅝ × ⅜ LLBB	33.5	2L 127 × 89 × 15.9 × 9 LLBB	48.9
2L 5 × 3½ × ⅝ × ¾ LLBB	33.5	2L 127 × 89 × 15.9 × 19 LLBB	48.9
2L 5 × 3½ × ½ LLBB	27.2	2L 127 × 89 × 12.7 LLBB	39.8
2L 5 × 3½ × ½ × ⅜ LLBB	27.2	2L 127 × 89 × 12.7 × 9 LLBB	39.8
2L 5 × 3½ × ½ × ¾ LLBB	27.2	2L 127 × 89 × 12.7 × 19 LLBB	39.8
2L 5 × 3½ × ⅜ LLBB	20.8	2L 127 × 89 × 9.5 LLBB	30.3
2L 5 × 3½ × ⅜ × ⅜ LLBB	20.8	2L 127 × 89 × 9.5 × 9 LLBB	30.3
2L 5 × 3½ × ⅜ × ¾ LLBB	20.8	2L 127 × 89 × 9.5 × 19 LLBB	30.3
2L 5 × 3½ × ⁵⁄₁₆ LLBB	17.4	2L 127 × 89 × 7.9 LLBB	25.4
2L 5 × 3½ × ⁵⁄₁₆ × ⅜ LLBB	17.4	2L 127 × 89 × 7.9 × 9 LLBB	25.4
2L 5 × 3½ × ⁵⁄₁₆ × ¾ LLBB	17.4	2L 127 × 89 × 7.9 × 19 LLBB	25.4
2L 5 × 3½ × ¼ LLBB	14.1	2L 127 × 89 × 6.4 LLBB	20.5
2L 5 × 3½ × ¼ × ⅜ LLBB	14.1	2L 127 × 89 × 6.4 × 9 LLBB	20.5
2L 5 × 3½ × ¼ × ¾ LLBB	14.1	2L 127 × 89 × 6.4 × 19 LLBB	20.5
2L 5 × 3 × ½ LLBB	25.5	2L 127 × 76 × 12.7 LLBB	37.3
2L 5 × 3 × ½ × ⅜ LLBB	25.5	2L 127 × 76 × 12.7 × 9 LLBB	37.3

STRUCTURAL STEEL WEIGHT TABLE (IMPERIAL AND METRIC) *(cont.)*

2L-LLBB (DOUBLE ANGLES WITH LONG LEGS BACK-TO-BACK)

Imperial		Metric	
Designation	Weight (lb/ft)	Designation	Weight (kg/m)
2L 5 × 3 × 1/2 × 3/4 LLBB	25.5	2L 127 × 76 × 12.7 × 19 LLBB	37.3
2L 5 × 3 × 7/16 LLBB	22.5	2L 127 × 76 × 11.1 LLBB	32.9
2L 5 × 3 × 7/16 × 3/8 LLBB	22.5	2L 127 × 76 × 11.1 × 9 LLBB	32.9
2L 5 × 3 × 7/16 × 3/4 LLBB	22.5	2L 127 × 76 × 11.1 × 19 LLBB	32.9
2L 5 × 3 × 3/8 LLBB	19.5	2L 127 × 76 × 9.5 LLBB	28.4
2L 5 × 3 × 3/8 × 3/8 LLBB	19.5	2L 127 × 76 × 9.5 × 9 LLBB	28.4
2L 5 × 3 × 3/8 × 3/4 LLBB	19.5	2L 127 × 76 × 9.5 × 19 LLBB	28.4
2L 5 × 3 × 5/16 LLBB	16.4	2L 127 × 76 × 7.9 LLBB	23.9
2L 5 × 3 × 5/16 × 3/8 LLBB	16.4	2L 127 × 76 × 7.9 × 9 LLBB	23.9
2L 5 × 3 × 5/16 × 3/4 LLBB	16.4	2L 127 × 76 × 7.9 × 19 LLBB	23.9
2L 5 × 3 × 1/4 LLBB	13.2	2L 127 × 76 × 6.4 LLBB	19.3
2L 5 × 3 × 1/4 × 3/8 LLBB	13.2	2L 127 × 76 × 6.4 × 9 LLBB	19.3
2L 5 × 3 × 1/4 × 3/4 LLBB	13.2	2L 127 × 76 × 6.4 × 19 LLBB	19.3
2L 4 × 3 1/2 × 1/2 LLBB	23.8	2L 102 × 89 × 12.7 LLBB	34.8
2L 4 × 3 1/2 × 1/2 × 3/8 LLBB	23.8	2L 102 × 89 × 12.7 × 9 LLBB	34.8
2L 4 × 3 1/2 × 1/2 × 3/4 LLBB	23.8	2L 102 × 89 × 12.7 × 19 LLBB	34.8
2L 4 × 3 1/2 × 3/8 LLBB	18.2	2L 102 × 89 × 9.5 LLBB	26.6
2L 4 × 3 1/2 × 3/8 × 3/8 LLBB	18.2	2L 102 × 89 × 9.5 × 9 LLBB	26.6
2L 4 × 3 1/2 × 3/8 × 3/4 LLBB	18.2	2L 102 × 89 × 9.5 × 19 LLBB	26.6
2L 4 × 3 1/2 × 5/16 LLBB	15.3	2L 102 × 89 × 7.9 LLBB	22.3
2L 4 × 3 1/2 × 5/16 × 3/8 LLBB	15.3	2L 102 × 89 × 7.9 × 9 LLBB	22.3
2L 4 × 3 1/2 × 5/16 × 3/4 LLBB	15.3	2L 102 × 89 × 7.9 × 19 LLBB	22.3

STRUCTURAL STEEL WEIGHT TABLE (IMPERIAL AND METRIC) *(cont.)*

2L-LLBB (DOUBLE ANGLES WITH LONG LEGS BACK-TO-BACK)

Imperial		Metric	
Designation	**Weight (lb/ft)**	**Designation**	**Weight (kg/m)**
2L 4 × 3½ × ¼ LLBB	12.4	2L 102 × 89 × 6.4 LLBB	18.0
2L 4 × 3½ × ¼ × ⅜ LLBB	12.4	2L 102 × 89 × 6.4 × 9 LLBB	18.0
2L 4 × 3½ × ¼ × ¾ LLBB	12.4	2L 102 × 89 × 6.4 × 19 LLBB	18.0
2L 4 × 3 × ⅝ LLBB	27.1	2L 102 × 76 × 15.9 LLBB	39.6
2L 4 × 3 × ⅝ × ⅜ LLBB	27.1	2L 102 × 76 × 15.9 × 9 LLBB	39.6
2L 4 × 3 × ⅝ × ¾ LLBB	27.1	2L 102 × 76 × 15.9 × 19 LLBB	39.6
2L 4 × 3 × ½ LLBB	22.1	2L 102 × 76 × 12.7 LLBB	32.3
2L 4 × 3 × ½ × ⅜ LLBB	22.1	2L 102 × 76 × 12.7 × 9 LLBB	32.3
2L 4 × 3 × ½ × 3/4 LLBB	22.1	2L 102 × 76 × 12.7 × 19 LLBB	32.3
2L 4 × 3 × ⅜ LLBB	16.9	2L 102 × 76 × 9.5 LLBB	24.7
2L 4 × 3 × ⅜ × ⅜ LLBB	16.9	2L 102 × 76 × 9.5 × 9 LLBB	24.7
2L 4 × 3 × ⅜ × ¾ LLBB	16.9	2L 102 × 76 × 9.5 × 19 LLBB	24.7
2L 4 × 3 × 5/16 LLBB	14.2	2L 102 × 76 × 7.9 LLBB	20.8
2L 4 × 3 × 5/16 × ⅜ LLBB	14.2	2L 102 × 76 × 7.9 × 9 LLBB	20.8
2L 4 × 3 × 5/16 × ¾ LLBB	14.2	2L 102 × 76 × 7.9 × 19 LLBB	20.8
2L 4 × 3 × ¼ LLBB	11.5	2L 102 × 76 × 6.4 LLBB	16.8
2L 4 × 3 × ¼ × ⅜ LLBB	11.5	2L 102 × 76 × 6.4 × 9 LLBB	16.8
2L 4 × 3 × ¼ × ¾ LLBB	11.5	2L 102 × 76 × 6.4 × 19 LLBB	16.8
2L 3½ × 3 × ½ LLBB	20.6	2L 89 × 76 × 12.7 LLBB	30.0
2L 3½ × 3 × ½ × ⅜ LLBB	20.6	2L 89 × 76 × 12.7 × 9 LLBB	30.0
2L 3½ × 3 × ½ × ¾ LLBB	20.6	2L 89 × 76 × 12.7 × 19 LLBB	30.0
2L 3½ × 3 × 7/16 LLBB	18.2	2L 89 × 76 × 11.1 LLBB	26.5

STRUCTURAL STEEL WEIGHT TABLE (IMPERIAL AND METRIC) *(cont.)*

2L-LLBB (DOUBLE ANGLES WITH LONG LEGS BACK-TO-BACK)

Imperial		Metric	
Designation	Weight (lb/ft)	Designation	Weight (kg/m)
2L 3½ × 3 × 7/16 × 3/8 LLBB	18.2	2L 89 × 76 × 11.1 × 9 LLBB	26.5
2L 3½ × 3 × 7/16 × 3/4 LLBB	18.2	2L 89 × 76 × 11.1 × 19 LLBB	26.5
2L 3½ × 3 × 3/8 LLBB	15.8	2L 89 × 76 × 9.5 LLBB	23.0
2L 3½ × 3 × 3/8 × 3/8 LLBB	15.8	2L 89 × 76 × 9.5 × 9 LLBB	23.0
2L 3½ × 3 × 3/8 × 3/4 LLBB	15.8	2L 89 × 76 × 9.5 × 19 LLBB	23.0
2L 3½ × 3 × 5/16 LLBB	13.3	2L 89 × 76 × 7.9 LLBB	19.4
2L 3½ × 3 × 5/16 × 3/8 LLBB	13.3	2L 89 × 76 × 7.9 × 9 LLBB	19.4
2L 3½ × 3 × 5/16 × 3/4 LLBB	13.3	2L 89 × 76 × 7.9 × 19 LLBB	19.4
2L 3½ × 3 × 1/4 LLBB	10.8	2L 89 × 76 × 6.4 LLBB	15.7
2L 3½ × 3 × 1/4 × 3/8 LLBB	10.8	2L 89 × 76 × 6.4 × 9 LLBB	15.7
2L 3½ × 3 × 1/4 × 3/4 LLBB	10.8	2L 89 × 76 × 6.4 × 19 LLBB	15.7
2L 3½ × 2½ × 1/2 LLBB	18.8	2L 89 × 64 × 12.7 LLBB	27.5
2L 3½ × 2½ × 1/2 × 3/8 LLBB	18.8	2L 89 × 64 × 12.7 × 9 LLBB	27.5
2L 3½ × 2½ × 1/2 × 3/4 LLBB	18.8	2L 89 × 64 × 12.7 × 19 LLBB	27.5
2L 3½ × 2½ × 3/8 LLBB	14.5	2L 89 × 64 × 9.5 LLBB	21.1
2L 3½ × 2½ × 3/8 × 3/8 LLBB	14.5	2L 89 × 64 × 9.5 × 9 LLBB	21.1
2L 3½ × 2½ × 3/8 × 3/4 LLBB	14.5	2L 89 × 64 × 9.5 × 19 LLBB	21.1
2L 3½ × 2½ × 5/16 LLBB	12.2	2L 89 × 64 × 7.9 LLBB	17.8
2L 3½ × 2½ × 5/16 × 3/8 LLBB	12.2	2L 89 × 64 × 7.9 × 9 LLBB	17.8
2L 3½ × 2½ × 5/16 × 3/4 LLBB	12.2	2L 89 × 64 × 7.9 × 19 LLBB	17.8
2L 3½ × 2½ × 1/4 LLBB	9.88	2L 89 × 64 × 6.4 LLBB	14.4
2L 3½ × 2½ × 1/4 × 3/8 LLBB	9.88	2L 89 × 64 × 6.4 × 9 LLBB	14.4

STRUCTURAL STEEL WEIGHT TABLE
(IMPERIAL AND METRIC) *(cont.)*

2L-LLBB (DOUBLE ANGLES WITH LONG LEGS BACK-TO-BACK)

Imperial		Metric	
Designation	Weight (lb/ft)	Designation	Weight (kg/m)
2L 3½ × 2½ × ¼ × ¾ LLBB	9.88	2L 89 × 64 × 6.4 × 19 LLBB	14.4
2L 3 × 2½ × ½ LLBB	17.1	2L 76 × 64 × 12.7 LLBB	24.9
2L 3 × 2½ × ½ × ⅜ LLBB	17.1	2L 76 × 64 × 12.7 × 9 LLBB	24.9
2L 2½ × ½ × ¾ LLBB	17.1	2L 76 × 64 × 12.7 × 19 LLBB	24.9
2L 3 × 2½ × ⁷⁄₁₆ LLBB	15.1	2L 76 × 64 × 11.1 LLBB	22.1
2L 3 × 2½ × ⁷⁄₁₆ × ⅜ LLBB	15.1	2L 76 × 64 × 11.1 × 9 LLBB	22.1
2L 3 × 2½ × ⁷⁄₁₆ × ¾ LLBB	15.1	2L 76 × 64 × 11.1 × 19 LLBB	22.1
2L 3 × 2½ × ⅜ LLBB	13.1	2L 76 × 64 × 9.5 LLBB	19.1
2L 3 × 2½ × ⅜ × ⅜ LLBB	13.1	2L 76 × 64 × 9.5 × 9 LLBB	19.1
2L 3 × 2½ × ⅜ × ¾ LLBB	13.1	2L 76 × 64 × 9.5 × 19 LLBB	19.1
2L 3 × 2½ × ⁵⁄₁₆ LLBB	11.1	2L 76 × 64 × 7.9 LLBB	16.2
2L 3 × 2½ × ⁵⁄₁₆ × ⅜ LLBB	11.1	2L 76 × 64 × 7.9 × 9 LLBB	16.2
2L 3 × 2½ × ⁵⁄₁₆ × ¾ LLBB	11.1	2L 76 × 64 × 7.9 × 19 LLBB	16.2
2L 3 × 2½ × ¼ LLBB	8.97	2L 76 × 64 × 6.4 LLBB	13.1
2L 3 × 2½ × ¼ × ⅜ LLBB	8.97	2L 76 × 64 × 6.4 × 9 LLBB	13.1
2L 3 × 2½ × ¼ × ¾ LLBB	8.97	2L 76 × 64 × 6.4 × 19 LLBB	13.1
2L 3 × 2½ × ³⁄₁₆ LLBB	6.82	2L 76 × 64 × 4.8 LLBB	10.0
2L 3 × 2½ × ³⁄₁₆ × ⅜ LLBB	6.82	2L 76 × 64 × 4.8 × 9 LLBB	10.0
2L 3 × 2½ × ³⁄₁₆ × ¾ LLBB	6.82	2L 76 × 64 × 4.8 × 19 LLBB	10.0
2L 3 × 2 × ½ LLBB	15.4	2L 76 × 51 × 12.7 LLBB	22.5
2L 3 × 2 × ½ × ⅜ LLBB	15.4	2L 76 × 51 × 12.7 × 9 LLBB	22.5
2L 3 × 2 × ½ × ¾ LLBB	15.4	2L 76 × 51 × 12.7 × 19 LLBB	22.5

STRUCTURAL STEEL WEIGHT TABLE (IMPERIAL AND METRIC) *(cont.)*

2L-LLBB (DOUBLE ANGLES WITH LONG LEGS BACK-TO-BACK)

Imperial		Metric	
Designation	**Weight (lb/ft)**	**Designation**	**Weight (kg/m)**
2L 3 × 2 × ⅜ LLBB	11.9	2L 76 × 51 × 9.5 LLBB	17.4
2L 3 × 2 × ⅜ × ⅜ LLBB	11.9	2L 76 × 51 × 9.5 × 9 LLBB	17.4
2L 3 × 2 × ⅜ × ¾ LLBB	11.9	2L 76 × 51 × 9.5 × 19 LLBB	17.4
2L 3 × 2 × ⁵⁄₁₆ LLBB	10.1	2L 76 × 51 × 7.9 LLBB	14.7
2L 3 × 2 × ⁵⁄₁₆ × ⅜ LLBB	10.1	2L 76 × 51 × 7.9 × 9 LLBB	14.7
2L 3 × 2 × ⁵⁄₁₆ × ¾ LLBB	10.1	2L 76 × 51 × 7.9 × 19 LLBB	14.7
2L 3 × 2 × ¼ LLBB	8.18	2L 76 × 51 × 6.4 LLBB	11.9
2L 3 × 2 × ¼ × ⅜ LLBB	8.18	2L 76 × 51 × 6.4 × 9 LLBB	11.9
2L 3 × 2 × ¼ × ¾ LLBB	8.18	2L 76 × 51 × 6.4 × 19 LLBB	11.9
2L 3 × 2 × ³⁄₁₆ LLBB	6.24	2L 76 × 51 × 4.8 LLBB	9.1
2L 3 × 2 × ³⁄₁₆ × ⅜ LLBB	6.24	2L 76 × 51 × 4.8 × 9 LLBB	9.1
2L 3 × 2 × ³⁄₁₆ × ¾ LLBB	6.24	2L 76 × 51 × 4.8 × 19 LLBB	9.1
2L 2½ × 2 × ⅜ LLBB	10.6	2L 64 × 51 × 9.5 LLBB	15.5
2L 2½ × 2 × ⅜ × ⅜ LLBB	10.6	2L 64 × 51 × 9.5 × 9 LLBB	15.5
2L 2½ × 2 × ⅜ × ¾ LLBB	10.6	2L 64 × 51 × 9.5 × 19 LLBB	15.5
2L 2½ × 2 × ⁵⁄₁₆ LLBB	8.97	2L 64 × 51 × 7.9 LLBB	13.1
2L 2½ × 2 × ⁵⁄₁₆ × ⅜ LLBB	8.97	2L 64 × 51 × 7.9 × 9 LLBB	13.1
2L 2½ × 2 × ⁵⁄₁₆ × ¾ LLBB	8.97	2L 64 × 51 × 7.9 × 19 LLBB	13.1
2L 2½ × 2 × ¼ LLBB	7.3	2L 64 × 51 × 6.4 LLBB	10.6
2L 2½ × 2 × ¼ × ⅜ LLBB	7.3	2L 64 × 51 × 6.4 × 9 LLBB	10.6
2L 2½ × 2 × ¼ × ¾ LLBB	7.3	2L 64 × 51 × 6.4 × 19 LLBB	10.6
2L 2½ × 2 × ³⁄₁₆ LLBB	5.57	2L 64 × 51 × 4.8 LLBB	8.1
2L 2½ × 2 × ³⁄₁₆ × ⅜ LLBB	5.57	2L 64 × 51 × 4.8 × 9 LLBB	8.1
2L 2½ × 2 × ³⁄₁₆ × ¾ LLBB	5.57	2L 64 × 51 × 4.8 × 19 LLBB	8.1

STRUCTURAL STEEL WEIGHT TABLE
(IMPERIAL AND METRIC) (cont.)

2L-SLBB (DOUBLE ANGLES WITH SHORT LEGS BACK-TO-BACK)

Imperial		Metric	
Designation	**Weight (lb/ft)**	**Designation**	**Weight (kg/m)**
2L 8 × 6 × 1 SLBB	88.8	2L 203 × 152 × 25.4 SLBB	130.0
2L 8 × 6 × 1 × 3/8 SLBB	88.8	2L 203 × 152 × 25.4 × 9 SLBB	130.0
2L 8 × 6 × 1 × 3/4 SLBB	88.8	2L 203 × 152 × 25.4 × 19 SLBB	130.0
2L 8 × 6 × 7/8 SLBB	78.5	2L 203 × 152 × 22.2 SLBB	115.0
2L 8 × 6 × 7/8 × 3/8 SLBB	78.5	2L 203 × 152 × 22.2 × 9 SLBB	115.0
2L 8 × 6 × 7/8 × 3/4 SLBB	78.5	2L 203 × 152 × 22.2 × 19 SLBB	115.0
2L 8 × 6 × 3/4 SLBB	68	2L 203 × 152 × 19 SLBB	99.2
2L 8 × 6 × 3/4 × 3/8 SLBB	68	2L 203 × 152 × 19 × 9 SLBB	99.2
2L 8 × 6 × 3/4 × 3/4 SLBB	68	2L 203 × 152 × 19 × 19 SLBB	99.2
2L 8 × 6 × 5/8 SLBB	57.3	2L 203 × 152 × 15.9 SLBB	83.6
2L 8 × 6 × 5/8 × 3/8 SLBB	57.3	2L 203 × 152 × 15.9 × 9 SLBB	83.6
2L 8 × 6 × 5/8 × 3/4 SLBB	57.3	2L 203 × 152 × 15.9 × 19 SLBB	83.6
2L 8 × 6 × 9/16 SLBB	51.8	2L 203 × 152 × 14.3 SLBB	75.6
2L 8 × 6 × 9/16 × 3/8 SLBB	51.8	2L 203 × 152 × 14.3 × 9 SLBB	75.6
2L 8 × 6 × 9/16 × 3/4 SLBB	51.8	2L 203 × 152 × 14.3 × 19 SLBB	75.6
2L 8 × 6 × 1/2 SLBB	46.3	2L 203 × 152 × 12.7 SLBB	67.6
2L 8 × 6 × 1/2 × 3/8 SLBB	46.3	2L 203 × 152 × 12.7 × 9 SLBB	67.6
2L 8 × 6 × 1/2 × 3/4 SLBB	46.3	2L 203 × 152 × 12.7 × 19 SLBB	67.6
2L 8 × 6 × 7/16 SLBB	40.7	2L 203 × 152 × 11.1 SLBB	59.5
2L 8 × 6 × 7/16 × 3/8 SLBB	40.7	2L 203 × 152 × 11.1 × 9 SLBB	59.5
2L 8 × 6 × 7/16 × 3/4 SLBB	40.7	2L 203 × 152 × 11.1 × 19 SLBB	59.5
2L 8 × 4 × 1 SLBB	75.2	2L 203 × 102 × 25.4 SLBB	110.0

STRUCTURAL STEEL WEIGHT TABLE (IMPERIAL AND METRIC) *(cont.)*

2L-SLBB (DOUBLE ANGLES WITH SHORT LEGS BACK-TO-BACK)

Imperial		Metric	
Designation	**Weight (lb/ft)**	**Designation**	**Weight (kg/m)**
2L 8 × 4 × 1 × ⅜ SLBB	75.2	2L 203 × 102 × 25.4 × 9 SLBB	110.0
2L 8 × 4 × 1 × ¾ SLBB	75.2	2L 203 × 102 × 25.4 × 19 SLBB	110.0
2L 8 × 4 × ⅞ SLBB	66.6	2L 203 × 102 × 22.2 SLBB	97.2
2L 8 × 4 × ⅞ × ⅜ SLBB	66.6	2L 203 × 102 × 22.2 × 9 SLBB	97.2
2L 8 × 4 × ⅞ × ¾ SLBB	66.6	2L 203 × 102 × 22.2 × 19 SLBB	97.2
2L 8 × 4 × ¾ SLBB	57.8	2L 203 × 102 × 19 SLBB	84.3
2L 8 × 4 × ¾ × ⅜ SLBB	57.8	2L 203 × 102 × 19 × 9 SLBB	84.3
2L 8 × 4 × ¾ × ¾ SLBB	57.8	2L 203 × 102 × 19 × 19 SLBB	84.3
2L 8 × 4 × ⅝ SLBB	48.7	2L 203 × 102 × 15.9 SLBB	71.1
2L 8 × 4 × ⅝ × ⅜ SLBB	48.7	2L 203 × 102 × 15.9 × 9 SLBB	71.1
2L 8 × 4 × ⅝ × ¾ SLBB	48.7	2L 203 × 102 × 15.9 × 19 SLBB	71.1
2L 8 × 4 × 9⁄16 SLBB	44.1	2L 203 × 102 × 14.3 SLBB	64.4
2L 8 × 4 × 9⁄16 × ⅜ SLBB	44.1	2L 203 × 102 × 14.3 × 9 SLBB	64.4
2L 8 × 4 × 9⁄16 × ¾ SLBB	44.1	2L 203 × 102 × 14.3 × 19 SLBB	64.4
2L 8 × 4 × ½ SLBB	39.5	2L 203 × 102 × 12.7 SLBB	57.6
2L 8 × 4 × ½ × ⅜ SLBB	39.5	2L 203 × 102 × 12.7 × 9 SLBB	57.6
2L 8 × 4 × ½ × ¾ SLBB	39.5	2L 203 × 102 × 12.7 × 19 SLBB	57.6
2L 8 × 4 × 7⁄16 SLBB	34.8	2L 203 × 102 × 11.1 SLBB	50.8
2L 8 × 4 × 7⁄16 × ⅜ SLBB	34.8	2L 203 × 102 × 11.1 × 9 SLBB	50.8
2L 8 × 4 × 7⁄16 × ¾ SLBB	34.8	2L 203 × 102 × 11.1 × 19 SLBB	50.8
2L 7 × 4 × ¾ SLBB	52.4	2L 178 × 102 × 19 SLBB	76.5
2L 7 × 4 × ¾ × ⅜ SLBB	52.4	2L 178 × 102 × 19 × 9 SLBB	76.5

STRUCTURAL STEEL WEIGHT TABLE (IMPERIAL AND METRIC) *(cont.)*

2L-SLBB (DOUBLE ANGLES WITH SHORT LEGS BACK-TO-BACK)

Imperial		Metric	
Designation	**Weight (lb/ft)**	**Designation**	**Weight (kg/m)**
2L 7 × 4 × ¾ × ¾ SLBB	52.4	2L 178 × 102 × 19 × 19 SLBB	76.5
2L 7 × 4 × ⅝ SLBB	44.2	2L 178 × 102 × 15.9 SLBB	64.5
2L 7 × 4 × ⅝ × ⅜ SLBB	44.2	2L 178 × 102 × 15.9 × 9 SLBB	64.5
2L 7 × 4 × ⅝ × ¾ SLBB	44.2	2L 178 × 102 × 15.9 × 19 SLBB	64.5
2L 7 × 4 × ½ SLBB	35.8	2L 178 × 102 × 12.7 SLBB	52.3
2L 7 × 4 × ½ × ⅜ SLBB	35.8	2L 178 × 102 × 12.7 × 9 SLBB	52.3
2L 7 × 4 × ½ × ¾ SLBB	35.8	2L 178 × 102 × 12.7 × 19 SLBB	52.3
2L 7 × 4 × ⁷⁄₁₆ SLBB	31.5	2L 178 × 102 × 11.1 SLBB	46.0
2L 7 × 4 × ⁷⁄₁₆ × ⅜ SLBB	31.5	2L 178 × 102 × 11.1 × 9 SLBB	46.0
2L 7 × 4 × ⁷⁄₁₆ × ¾ SLBB	31.5	2L 178 × 102 × 11.1 × 19 SLBB	46.0
2L 7 × 4 × ⅜ SLBB	27.2	2L 178 × 102 × 9.5 SLBB	39.7
2L 7 × 4 × ⅜ × ⅜ SLBB	27.2	2L 178 × 102 × 9.5 × 9 SLBB	39.7
2L 7 × 4 × ⅜ × ¾ SLBB	27.2	2L 178 × 102 × 9.5 × 19 SLBB	39.7
2L 6 × 4 × ⅞ SLBB	54.3	2L 152 × 102 × 22.2 SLBB	79.3
2L 6 × 4 × ⅞ × ⅜ SLBB	54.3	2L 152 × 102 × 22.2 × 9 SLBB	79.3
2L 6 × 4 × ⅞ × ¾ SLBB	54.3	2L 152 × 102 × 22.2 × 19 SLBB	79.3
2L 6 × 4 × ¾ SLBB	47.2	2L 152 × 102 × 19 SLBB	68.9
2L 6 × 4 × ¾ × ⅜ SLBB	47.2	2L 152 × 102 × 19 × 9 SLBB	68.9
2L 6 × 4 × ¾ × ¾ SLBB	47.2	2L 152 × 102 × 19 × 19 SLBB	68.9
2L 6 × 4 × ⅝ SLBB	39.9	2L 152 × 102 × 15.9 SLBB	58.2
2L 6 × 4 × ⅝ × ⅜ SLBB	39.9	2L 152 × 102 × 15.9 × 9 SLBB	58.2
2L 6 × 4 × ⅝ × ¾ SLBB	39.9	2L 152 × 102 × 15.9 × 19 SLBB	58.2

STRUCTURAL STEEL WEIGHT TABLE (IMPERIAL AND METRIC) *(cont.)*

2L-SLBB (DOUBLE ANGLES WITH SHORT LEGS BACK-TO-BACK)

Imperial		Metric	
Designation	**Weight (lb/ft)**	**Designation**	**Weight (kg/m)**
2L 6 × 4 × $^9/_{16}$ SLBB	36.1	2L 152 × 102 × 14.3 SLBB	52.7
2L 6 × 4 × $^9/_{16}$ × $^3/_8$ SLBB	36.1	2L 152 × 102 × 14.3 × 9 SLBB	52.7
2L 6 × 4 × $^9/_{16}$ × $^3/_4$ SLBB	36.1	2L 152 × 102 × 14.3 × 19 SLBB	52.7
2L 6 × 4 × $^1/_2$ SLBB	32.3	2L 152 × 102 × 12.7 SLBB	47.2
2L 6 × 4 × $^1/_2$ × $^3/_8$ SLBB	32.3	2L 152 × 102 × 12.7 × 9 SLBB	47.2
2L 6 × 4 × $^1/_2$ × $^3/_4$ SLBB	32.3	2L 152 × 102 × 12.7 × 19 SLBB	47.2
2L 6 × 4 × $^7/_{16}$ SLBB	28.5	2L 152 × 102 × 11.1 SLBB	41.5
2L 6 × 4 × $^7/_{16}$ × $^3/_8$ SLBB	28.5	2L 152 × 102 × 11.1 × 9 SLBB	41.5
2L 6 × 4 × $^7/_{16}$ × $^3/_4$ SLBB	28.5	2L 152 × 102 × 11.1 × 19 SLBB	41.5
2L 6 × 4 × $^3/_8$ SLBB	24.6	2L 152 × 102 × 9.5 SLBB	35.8
2L 6 × 4 × $^3/_8$ × $^3/_8$ SLBB	24.6	2L 152 × 102 × 9.5 × 9 SLBB	35.8
2L 6 × 4 × $^3/_8$ × $^3/_4$ SLBB	24.6	2L 152 × 102 × 9.5 × 19 SLBB	35.8
2L 6 × 4 × $^5/_{16}$ SLBB	20.6	2L 152 × 102 × 7.9 SLBB	30.1
2L 6 × 4 × $^5/_{16}$ × $^3/_8$ SLBB	20.6	2L 152 × 102 × 7.9 × 9 SLBB	30.1
2L 6 × 4 × $^5/_{16}$ × $^3/_4$ SLBB	20.6	2L 152 × 102 × 7.9 × 19 SLBB	30.1
2L 6 × 3$^1/_2$ × $^1/_2$ SLBB	30.7	2L 152 × 89 × 12.7 SLBB	44.9
2L 6 × 3$^1/_2$ × $^1/_2$ × $^3/_8$ SLBB	30.7	2L 152 × 89 × 12.7 × 9 SLBB	44.9
2L 6 × 3$^1/_2$ × $^1/_2$ × $^3/_4$ SLBB	30.7	2L 152 × 89 × 12.7 × 19 SLBB	44.9
2L 6 × 3$^1/_2$ × $^3/_8$ SLBB	23.4	2L 152 × 89 × 9.5 SLBB	34.2
2L 6 × 3$^1/_2$ × $^3/_8$ × $^3/_8$ SLBB	23.4	2L 152 × 89 × 9.5 × 9 SLBB	34.2
2L 6 × 3$^1/_2$ × $^3/_8$ × $^3/_4$ SLBB	23.4	2L 152 × 89 × 9.5 × 19 SLBB	34.2
2L 6 × 3$^1/_2$ × $^5/_{16}$ SLBB	19.7	2L 152 × 89 × 7.9 SLBB	28.7

STRUCTURAL STEEL WEIGHT TABLE (IMPERIAL AND METRIC) *(cont.)*

2L-SLBB (DOUBLE ANGLES WITH SHORT LEGS BACK-TO-BACK)

Imperial		Metric	
Designation	**Weight (lb/ft)**	**Designation**	**Weight (kg/m)**
2L 6 × 3½ × ⁵⁄₁₆ × ⅜ SLBB	19.7	2L 152 × 89 × 7.9 × 9 SLBB	28.7
2L 6 × 3½ × ⁵⁄₁₆ × ¾ SLBB	19.7	2L 152 × 89 × 7.9 × 19 SLBB	28.7
2L 5 × 3½ × ¾ SLBB	39.6	2L 127 × 89 × 19 SLBB	57.8
2L 5 × 3½ × ¾ × ⅜ SLBB	39.6	2L 127 × 89 × 19 × 9 SLBB	57.8
2L 5 × 3½ × ¾ × ¾ SLBB	39.6	2L 127 × 89 × 19 × 19 SLBB	57.8
2L 5 × 33½ × ⅝ SLBB	33.5	2L 127 × 89 × 15.9 SLBB	48.9
2L 5 × 3½ × ⅝ × ⅜ SLBB	33.5	2L 127 × 89 × 15.9 × 9 SLBB	48.9
2L 5 × 3½ × ⅝ × ¾ SLBB	33.5	2L 127 × 89 × 15.9 × 19 SLBB	48.9
2L 5 × 3½ × ½ SLBB	27.2	2L 127 × 89 × 12.7 SLBB	39.8
2L 5 × 3½ × ½ × ⅜ SLBB	27.2	2L 127 × 89 × 12.7 × 9 SLBB	39.8
2L 5 × 3½ × ½ × ¾ SLBB	27.2	2L 127 × 89 × 12.7 × 19 SLBB	39.8
2L 5 × 3½ × ⅜ SLBB	20.8	2L 127 × 89 × 9.5 SLBB	30.3
2L 5 × 3½ × ⅜ × ⅜ SLBB	20.8	2L 127 × 89 × 9.5 × 9 SLBB	30.3
2L 5 × 3½ × ⅜ × ¾ SLBB	20.8	2L 127 × 89 × 9.5 × 19 SLBB	30.3
2L 5 × 3½ × ⁵⁄₁₆ SLBB	17.4	2L 127 × 89 × 7.9 SLBB	25.4
2L 5 × 3½ × ⁵⁄₁₆ × ⅜ SLBB	17.4	2L 127 × 89 × 7.9 × 9 SLBB	25.4
2L 5 × 3½ × ⁵⁄₁₆ × ¾ SLBB	17.4	2L 127 × 89 × 7.9 × 19 SLBB	25.4
2L 5 × 3½ × ¼ SLBB	14.1	2L 127 × 89 × 6.4 SLBB	20.5
2L 5 × 3½ × ¼ × ⅜ SLBB	14.1	2L 127 × 89 × 6.4 × 9 SLBB	20.5
2L 5 × 3½ × ¼ × ¾ SLBB	14.1	2L 127 × 89 × 6.4 × 19 SLBB	20.5
2L 5 × 3 × ½ SLBB	25.5	2L 127 × 76 × 12.7 SLBB	37.3
2L 5 × 3 × ½ × ⅜ SLBB	25.5	2L 127 × 76 × 12.7 × 9 SLBB	37.3

STRUCTURAL STEEL WEIGHT TABLE (IMPERIAL AND METRIC) *(cont.)*

2L-SLBB (DOUBLE ANGLES WITH SHORT LEGS BACK-TO-BACK)

Imperial		Metric	
Designation	**Weight (lb/ft)**	**Designation**	**Weight (kg/m)**
2L 5 × 3 × ½ × ¾ SLBB	25.5	2L 127 × 76 × 12.7 × 19 SLBB	37.3
2L 5 × 3 × $^7/_{16}$ SLBB	22.5	2L 127 × 76 × 11.1 SLBB	32.9
2L 5 × 3 × $^7/_{16}$ × ⅜ SLBB	22.5	2L 127 × 76 × 11.1 × 9 SLBB	32.9
2L 5 × 3 × $^7/_{16}$ × ¾ SLBB	22.5	2L 127 × 76 × 11.1 × 19 SLBB	32.9
2L 5 × 3 × ⅜ SLBB	19.5	2L 127 × 76 × 9.5 SLBB	28.4
2L 5 × 3 × ⅜ × ⅜ SLBB	19.5	2L 127 × 76 × 9.5 × 9 SLBB	28.4
2L 5 × 3 × ⅜ × ¾ SLBB	19.5	2L 127 × 76 × 9.5 × 19 SLBB	28.4
2L 5 × 3 × $^5/_{16}$ SLBB	16.4	2L 127 × 76 × 7.9 SLBB	23.9
2L 5 × 3 × $^5/_{16}$ × ⅜ SLBB	16.4	2L 127 × 76 × 7.9 × 9 SLBB	23.9
2L 5 × 3 × $^5/_{16}$ × ¾ SLBB	16.4	2L 127 × 76 × 7.9 × 19 SLBB	23.9
2L 5 × 3 × ¼ SLBB	13.2	2L 127 × 76 × 6.4 SLBB	19.3
2L 5 × 3 × ¼ × ⅜ SLBB	13.2	2L 127 × 76 × 6.4 × 9 SLBB	19.3
2L 5 × 3 × ¼ × ¾ SLBB	13.2	2L 127 × 76 × 6.4 × 19 SLBB	19.3
2L 4 × 3½ × ½ SLBB	23.8	2L 102 × 89 × 12.7 SLBB	34.8
2L 4 × 3½ × ½ × ⅜ SLBB	23.8	2L 102 × 89 × 12.7 × 9 SLBB	34.8
2L 4 × 3½ × ½ × ¾ SLBB	23.8	2L 102 × 89 × 12.7 × 19 SLBB	34.8
2L 4 × 3½ × ⅜ SLBB	18.2	2L 102 × 89 × 9.5 SLBB	26.6
2L 4 × 3½ × ⅜ × ⅜ SLBB	18.2	2L 102 × 89 × 9.5 × 9 SLBB	26.6
2L 4 × 3½ × ⅜ × ¾ SLBB	18.2	2L 102 × 89 × 9.5 × 19 SLBB	26.6
2L 4 × 3½ × $^5/_{16}$ SLBB	15.3	2L 102 × 89 × 7.9 SLBB	22.3
2L 4 × 3½ × $^5/_{16}$ × ⅜ SLBB	15.3	2L 102 × 89 × 7.9 × 9 SLBB	22.3
2L 4 × 3½ × $^5/_{16}$ × ¾ SLBB	15.3	2L 102 × 89 × 7.9 × 19 SLBB	22.3

STRUCTURAL STEEL WEIGHT TABLE
(IMPERIAL AND METRIC) *(cont.)*

2L-SLBB (DOUBLE ANGLES WITH SHORT LEGS BACK-TO-BACK)

Imperial		Metric	
Designation	**Weight (lb/ft)**	**Designation**	**Weight (kg/m)**
2L 4 × 3½ × ¼ SLBB	12.4	2L 102 × 89 × 6.4 SLBB	18.0
2L 4 × 3½ × ¼ × ⅜ SLBB	12.4	2L 102 × 89 × 6.4 × 9 SLBB	18.0
2L 4 × 3½ × ¼ × ¾ SLBB	12.4	2L 102 × 89 × 6.4 × 19 SLBB	18.0
2L 4 × 3 × ⅝ SLBB	27.1	2L 102 × 76 × 15.9 SLBB	39.6
2L 4 × 3 × ⅝ × ⅜ SLBB	27.1	2L 102 × 76 × 15.9 × 9 SLBB	39.6
2L 4 × 3 × ⅝ × ¾ SLBB	27.1	2L 102 × 76 × 15.9 × 19 SLBB	39.6
2L 4 × 3 × ½ SLBB	22.1	2L 102 × 76 × 12.7 SLBB	32.3
2L 4 × 3 × ½ × ⅜ SLBB	22.1	2L 102 × 76 × 12.7 × 9 SLBB	32.3
2L 4 × 3 × ½ × ¾ SLBB	22.1	2L 102 × 76 × 12.7 × 19 SLBB	32.3
2L 4 × 3 × ⅜ SLBB	16.9	2L 102 × 76 × 9.5 SLBB	24.7
2L 4 × 3 × ⅜ × ⅜ SLBB	16.9	2L 102 × 76 × 9.5 × 9 SLBB	24.7
2L 4 × 3 × ⅜ × ¾ SLBB	16.9	2L 102 × 76 × 9.5 × 19 SLBB	24.7
2L 4 × 3 × ⁵⁄₁₆ SLBB	14.2	2L 102 × 76 × 7.9 SLBB	20.8
2L 4 × 3 × ⁵⁄₁₆ × ⅜ SLBB	14.2	2L 102 × 76 × 7.9 × 9 SLBB	20.8
2L 4 × 3 × ⁵⁄₁₆ × ¾ SLBB	14.2	2L 102 × 76 × 7.9 × 19 SLBB	20.8
2L 4 × 3 × ¼ SLBB	11.5	2L 102 × 76 × 6.4 SLBB	16.8
2L 4 × 3 × ¼ × ⅜ SLBB	11.5	2L 102 × 76 × 6.4 × 9 SLBB	16.8
2L 4 × 3 × ¼ × ¾ SLBB	11.5	2L 102 × 76 × 6.4 × 19 SLBB	16.8
2L 3½ × 3 × ½ SLBB	20.6	2L 89 × 76 × 12.7 SLBB	30.0
2L 3½ × 3 × ½ × ⅜ SLBB	20.6	2L 89 × 76 × 12.7 × 9 SLBB	30.0
2L 3½ × 3 × ½ × ¾ SLBB	20.6	2L 89 × 76 × 12.7 × 19 SLBB	30.0
2L 3½ × 3 × ⁷⁄₁₆ SLBB	18.2	2L 89 × 76 × 11.1 SLBB	26.5

STRUCTURAL STEEL WEIGHT TABLE (IMPERIAL AND METRIC) *(cont.)*

2L-SLBB (DOUBLE ANGLES WITH SHORT LEGS BACK-TO-BACK)

Imperial		Metric	
Designation	Weight (lb/ft)	Designation	Weight (kg/m)
2L 3½ × 3 × 7⁄16 × 3⁄8 SLBB	18.2	2L 89 × 76 × 11.1 × 9 SLBB	26.5
2L 3½ × 3 × 7⁄16 × 3⁄4 SLBB	18.2	2L 89 × 76 × 11.1 × 19 SLBB	26.5
2L 3½ × 3 × 3⁄8 SLBB	15.8	2L 89 × 76 × 9.5 SLBB	23.0
2L 3½ × 3 × 3⁄8 × 3⁄8 SLBB	15.8	2L 89 × 76 × 9.5 × 9 SLBB	23.0
2L 3½ × 3 × 3⁄8 × 3⁄4 SLBB	15.8	2L 89 × 76 × 9.5 × 19 SLBB	23.0
2L 3½ × 3 × 5⁄16 SLBB	13.3	2L 89 × 76 × 7.9 SLBB	19.4
2L 3½ × 3 × 5⁄16 × 3⁄8 SLBB	13.3	2L 89 × 76 × 7.9 × 9 SLBB	19.4
2L 3½ × 3 × 5⁄16 × 3⁄4 SLBB	13.3	2L 89 × 76 × 7.9 × 19 SLBB	19.4
2L 3½ × 3 × 1⁄4 SLBB	10.8	2L 89 × 76 × 6.4 SLBB	15.7
2L 3½ × 3 × 1⁄4 × 3⁄8 SLBB	10.8	2L 89 × 76 × 6.4 × 9 SLBB	15.7
2L 3½ × 3 × 1⁄4 × 3⁄4 SLBB	10.8	2L 89 × 76 × 6.4 × 19 SLBB	15.7
2L 3½ × 2½ × 1⁄2 SLBB	18.8	2L 89 × 64 × 12.7 SLBB	27.5
2L 3½ × 2½ × 1⁄2 × 3⁄8 SLBB	18.8	2L 89 × 64 × 12.7 × 9 SLBB	27.5
2L 3½ × 2½ × 1⁄2 × 3⁄4 SLBB	18.8	2L 89 × 64 × 12.7 × 19 SLBB	27.5
2L 3½ × 2½ × 3⁄8 SLBB	14.5	2L 89 × 64 × 9.5 SLBB	21.1
2L 3½ × 2½ × 3⁄8 × 3⁄8 SLBB	14.5	2L 89 × 64 × 9.5 × 9 SLBB	21.1
2L 3½ × 2½ × 3⁄8 × 3⁄4 SLBB	14.5	2L 89 × 64 × 9.5 × 19 SLBB	21.1
2L 3½ × 2½ × 5⁄16 SLBB	12.2	2L 89 × 64 × 7.9 SLBB	17.8
2L 3½ × 2½ × 5⁄16 × 3⁄8 SLBB	12.2	2L 89 × 64 × 7.9 × 9 SLBB	17.8
2L 3½ × 2½ × 5⁄16 × 3⁄4 SLBB	12.2	2L 89 × 64 × 7.9 × 19 SLBB	17.8
2L 3½ × 2½ × 1⁄4 SLBB	9.88	2L 89 × 64 × 6.4 SLBB	14.4
2L 3½ × 2½ × 1⁄4 × 3⁄8 SLBB	9.88	2L 89 × 64 × 6.4 × 9 SLBB	14.4

STRUCTURAL STEEL WEIGHT TABLE (IMPERIAL AND METRIC) *(cont.)*

2L-SLBB (DOUBLE ANGLES WITH SHORT LEGS BACK-TO-BACK)

Imperial		Metric	
Designation	Weight (lb/ft)	Designation	Weight (kg/m)
2L 3½ × 2½ × ¼ × ¾ SLBB	9.88	2L 89 × 64 × 6.4 × 19 SLBB	14.4
2L 3 × 2½ × ½ SLBB	17.1	2L 76 × 64 × 12.7 SLBB	24.9
2L 3 × 2½ × ½ × ⅜ SLBB	17.1	2L 76 × 64 × 12.7 × 9 SLBB	24.9
2L 3 × 2½ × ½ × ¾ SLBB	17.1	2L 76 × 64 × 12.7 × 19 SLBB	24.9
2L 3 × 2½ × ⁷⁄₁₆ SLBB	15.1	2L 76 × 64 × 11.1 SLBB	22.1
2L 3 × 2½ × ⁷⁄₁₆ × ⅜ SLBB	15.1	2L 76 × 64 × 11.1 × 9 SLBB	22.1
2L 3 × 2½ × ⁷⁄₁₆ × ¾ SLBB	15.1	2L 76 × 64 × 11.1 × 19 SLBB	22.1
2L 3 × 2½ × ⅜ SLBB	13.1	2L 76 × 64 × 9.5 SLBB	19.1
2L 3 × 2½ × ⅜ × ⅜ SLBB	13.1	2L 76 × 64 × 9.5 × 9 SLBB	19.1
2L 3 × 2½ × ⅜ × ¾ SLBB	13.1	2L 76 × 64 × 9.5 × 19 SLBB	19.1
2L 3 × 2½ × ⁵⁄₁₆ SLBB	11.1	2L 76 × 64 × 7.9 SLBB	16.2
2L 3 × 2½ × ⁵⁄₁₆ × ⅜ SLBB	11.1	2L 76 × 64 × 7.9 × 9 SLBB	16.2
2L 3 × 2½ × ⁵⁄₁₆ × ¾ SLBB	11.1	2L 76 × 64 × 7.9 × 19 SLBB	16.2
2L 3 × 2½ × ¼ SLBB	8.97	2L 76 × 64 × 6.4 SLBB	13.1
2L 3 × 2½ × ¼ × ⅜ SLBB	8.97	2L 76 × 64 × 6.4 × 9 SLBB	13.1
2L 3 × 2½ × ¼ × ¾ SLBB	8.97	2L 76 × 64 × 6.4 × 19 SLBB	13.1
2L 3 × 2½ × ³⁄₁₆ SLBB	6.82	2L 76 × 64 × 4.8 SLBB	10.0
2L 3 × 2½ × ³⁄₁₆ × ⅜ SLBB	6.82	2L 76 × 64 × 4.8 × 9 SLBB	10.0
2L 3 × 2½ × ³⁄₁₆ × ¾ SLBB	6.82	2L 76 × 64 × 4.8 × 19 SLBB	10.0
2L 3 × 2 × ½ SLBB	15.4	2L 76 × 51 × 12.7 SLBB	22.5
2L 3 × 2 × ½ × ⅜ SLBB	15.4	2L 76 × 51 × 12.7 × 9 SLBB	22.5
2L 3 × 2 × ½ × ¾ SLBB	15.4	2L 76 × 51 × 12.7 × 19 SLBB	22.5

STRUCTURAL STEEL WEIGHT TABLE (IMPERIAL AND METRIC) *(cont.)*

2L-SLBB (DOUBLE ANGLES WITH SHORT LEGS BACK-TO-BACK)

Imperial		Metric	
Designation	Weight (lb/ft)	Designation	Weight (kg/m)
2L 3 × 2 × ⅜ SLBB	11.9	2L 76 × 51 × 9.5 SLBB	17.4
2L 3 × 2 × ⅜ × 3/8 SLBB	11.9	2L 76 × 51 × 9.5 × 9 SLBB	17.4
2L 3 × 2 × ⅜ × ¾ SLBB	11.9	2L 76 × 51 × 9.5 × 19 SLBB	17.4
2L 3 × 2 × ⁵⁄₁₆ SLBB	10.1	2L 76 × 51 × 7.9 SLBB	14.7
2L 3 × 2 × ⁵⁄₁₆ × ⅜ SLBB	10.1	2L 76 × 51 × 7.9 × 9 SLBB	14.7
2L 3 × 2 × ⁵⁄₁₆ × ¾ SLBB	10.1	2L 76 × 51 × 7.9 × 19 SLBB	14.7
2L 3 × 2 × ¼ SLBB	8.18	2L 76 × 51 × 6.4 SLBB	11.9
2L 3 × 2 × ¼ × ⅜ SLBB	8.18	2L 76 × 51 × 6.4 × 9 SLBB	11.9
2L 3 × 2 × ¼ × ¾ SLBB	8.18	2L 76 × 51 × 6.4 × 19 SLBB	11.9
2L 3 × 2 × ³⁄₁₆ SLBB	6.24	2L 76 × 51 × 4.8 SLBB	9.1
2L 3 × 2 × ³⁄₁₆ × ⅜ SLBB	6.24	2L 76 × 51 × 4.8 × 9 SLBB	9.1
2L 3 × 2 × ³⁄₁₆ × ¾ SLBB	6.24	2L 76 × 51 × 4.8 × 19 SLBB	9.1
2L 2½ × 2 × ⅜ SLBB	10.6	2L 64 × 51 × 9.5 SLBB	15.5
2L 2½ × 2 × ⅜ × ⅜ SLBB	10.6	2L 64 × 51 × 9.5 × 9 SLBB	15.5
2L 2½ × 2 × ⅜ × ¾ SLBB	10.6	2L 64 × 51 × 9.5 × 19 SLBB	15.5
2L 2½ × 2 × ⁵⁄₁₆ SLBB	8.97	2L 64 × 51 × 7.9 SLBB	13.1
2L 2½ × 2 × ⁵⁄₁₆ × ⅜ SLBB	8.97	2L 64 × 51 × 7.9 × 9 SLBB	13.1
2L 2½ × 2 × ⁵⁄₁₆ × ¾ SLBB	8.97	2L 64 × 51 × 7.9 × 19 SLBB	13.1
2L 2½ × 2 × ¼ SLBB	7.3	2L 64 × 51 × 6.4 SLBB	10.6
2L 2½ × 2 × ¼ × ⅜ SLBB	7.3	2L 64 × 51 × 6.4 × 9 SLBB	10.6
2L 2½ × 2 × ¼ × ¾ SLBB	7.3	2L 64 × 51 × 6.4 × 19 SLBB	10.6
2L 2½ × 2 × ³⁄₁₆ SLBB	5.57	2L 64 × 51 × 4.8 SLBB	8.1
2L 2½ × 2 × ³⁄₁₆ × ⅜ SLBB	5.57	2L 64 × 51 × 4.8 × 9 SLBB	8.1
2L 2½ × 2 × ³⁄₁₆ × ¾ SLBB	5.57	2L 64 × 51 × 4.8 × 19 SLBB	8.1

STRUCTURAL STEEL WEIGHT TABLE
(IMPERIAL AND METRIC) *(cont.)*

HSS (HOLLOW STRUCTURAL SECTIONS) RECTANGULAR

Imperial		Metric	
Designation	**Weight (lb/ft)**	**Designation**	**Weight (kg/m)**
HSS 32 × 24 × 5/8	225.80	HSS 812.8 × 609.6 × 15.9	336.6
HSS 32 × 24 × 1/2	183.50	HSS 812.8 × 609.6 × 12.7	273.1
HSS 32 × 24 × 3/8	138.95	HSS 812.8 × 609.6 × 9.5	206.3
HSS 30 × 24 × 5/8	217.30	HSS 762.0 × 609.6 × 15.9	323.9
HSS 30 × 24 × 1/2	176.70	HSS 762.0 × 609.6 × 12.7	263.0
HSS 30 × 24 × 3/8	133.84	HSS 762.0 × 609.6 × 9.5	198.7
HSS 28 × 24 × 5/8	208.79	HSS 711.2 × 609.6 × 15.9	311.2
HSS 28 × 24 × 1/2	169.89	HSS 711.2 × 609.6 × 12.7	252.9
HSS 28 × 24 × 3/8	128.74	HSS 711.2 × 609.6 × 9.5	191.1
HSS 26 × 24 × 5/8	200.28	HSS 660.4 × 609.6 × 15.9	298.5
HSS 26 × 24 × 1/2	163.08	HSS 660.4 × 609.6 × 12.7	242.7
HSS 26 × 24 × 3/8	123.64	HSS 660.4 × 609.6 × 9.5	183.5
HSS 24 × 22 × 5/8	183.27	HSS 609.6 × 558.8 × 15.9	273.2
HSS 24 × 22 × 1/2	149.47	HSS 609.6 × 558.8 × 12.7	222.5
HSS 24 × 22 × 3/8	113.43	HSS 609.6 × 558.8 × 9.5	168.4
HSS 22 × 20 × 5/8	166.25	HSS 558.8 × 508.0 × 15.9	247.8
HSS 22 × 20 × 1/2	135.86	HSS 558.8 × 508.0 × 12.7	202.2
HSS 22 × 20 × 3/8	103.22	HSS 558.8 × 508.0 × 9.5	153.2

STRUCTURAL STEEL WEIGHT TABLE
(IMPERIAL AND METRIC) *(cont.)*

HSS (HOLLOW STRUCTURAL SECTIONS) RECTANGULAR

Imperial		Metric	
Designation	**Weight (lb/ft)**	**Designation**	**Weight (kg/m)**
HSS 20 × 18 × 5/8	149.24	HSS 508.0 × 457.2 × 15.9	222.4
HSS 20 × 18 × 1/2	122.25	HSS 508.0 × 457.2 × 12.7	182.0
HSS 20 × 18 × 3/8	93.01	HSS 508.0 × 457.2 × 9.5	138.1
HSS 20 × 16 × 5/8	140.73	HSS 508.0 × 406.4 × 15.9	209.8
HSS 20 × 16 × 1/2	115.45	HSS 508.0 × 406.4 × 12.7	171.8
HSS 20 × 16 × 3/8	87.91	HSS 508.0 × 406.4 × 9.5	130.5
HSS 20 × 12 × 5/8	123.72	HSS 508.0 × 304.8 × 15.9	184.4
HSS 20 × 12 × 1/2	103.30	HSS 508.0 × 304.8 × 12.7	153.7
HSS 20 × 12 × 3/8	78.52	HSS 508.0 × 304.8 × 9.5	116.6
HSS 20 × 12 × 5/16	65.87	HSS 508.0 × 304.8 × 7.9	97.6
HSS 20 × 8 × 5/8	110.36	HSS 508.0 × 203.2 × 15.9	164.5
HSS 20 × 8 × 1/2	89.68	HSS 508.0 × 203.2 × 12.7	133.5
HSS 20 × 8 × 3/8	68.31	HSS 508.0 × 203.2 × 9.5	101.4
HSS 20 × 8 × 5/16	57.36	HSS 508.0 × 203.2 × 7.9	85.0
HSS 20 × 4 × 1/2	76.07	HSS 508.0 × 101.6 × 12.7	113.2
HSS 20 × 4 × 3/8	58.10	HSS 508.0 × 101.6 × 9.5	86.3
HSS 20 × 4 × 5/16	48.86	HSS 508.0 × 101.6 × 7.9	72.4
HSS 18 × 12 × 5/8	115.21	HSS 457.2 × 304.8 × 15.9	171.7

STRUCTURAL STEEL WEIGHT TABLE (IMPERIAL AND METRIC) *(cont.)*

HSS (HOLLOW STRUCTURAL SECTIONS) RECTANGULAR

Imperial		Metric	
Designation	**Weight (lb/ft)**	**Designation**	**Weight (kg/m)**
HSS 18 × 12 × 1/2	95.03	HSS 457.2 × 304.8 × 12.7	141.4
HSS 18 × 12 × 3/8	72.59	HSS 457.2 × 304.8 × 9.5	107.8
HSS 18 × 6 × 5/8	93.34	HSS 457.2 × 152.4 × 15.9	139.1
HSS 18 × 6 × 1/2	76.07	HSS 457.2 × 152.4 × 12.7	113.2
HSS 18 × 6 × 3/8	58.10	HSS 457.2 × 152.4 × 9.5	86.3
HSS 18 × 6 × 5/16	48.86	HSS 457.2 × 152.4 × 7.9	72.4
HSS 18 × 6 × 1/4	39.43	HSS 457.2 × 152.4 × 6.4	59.1
HSS 16 × 12 × 5/8	106.71	HSS 406.4 × 304.8 × 15.9	159.0
HSS 16 × 12 × 1/2	89.68	HSS 406.4 × 304.8 × 12.7	133.5
HSS 16 × 12 × 3/8	68.31	HSS 406.4 × 304.8 × 9.5	101.4
HSS 16 × 12 × 5/16	57.36	HSS 406.4 × 304.8 × 7.9	85.0
HSS 16 × 8 × 5/8	93.34	HSS 406.4 × 203.2 × 15.9	139.1
HSS 16 × 8 × 1/2	76.07	HSS 406.4 × 304.8 × 12.7	113.2
HSS 16 × 8 × 3/8	58.10	HSS 406.4 × 304.8 × 9.5	86.3
HSS 16 × 8 × 5/16	48.86	HSS 406.4 × 304.8 × 7.9	72.4
HSS 16 × 4 × 1/2	62.46	HSS 406.4 × 101.6 × 12.7	93.0
HSS 16 × 4 × 3/8	47.90	HSS 406.4 × 304.8 × 9.5	71.1
HSS 16 × 4 × 5/16	40.35	HSS 406.4 × 304.8 × 7.9	59.8

STRUCTURAL STEEL WEIGHT TABLE (IMPERIAL AND METRIC) *(cont.)*

HSS (HOLLOW STRUCTURAL SECTIONS) RECTANGULAR

Imperial		Metric	
Designation	**Weight (lb/ft)**	**Designation**	**Weight (kg/m)**
HSS 14 × 12 × 1/2	81.42	HSS 355.6 × 304.8 × 12.7	121.2
HSS 14 × 12 × 3/8	62.39	HSS 355.6 × 304.8 × 9.5	92.6
HSS 14 × 10 × 5/8	93.34	HSS 355.6 × 254.0 × 15.9	139.1
HSS 14 × 10 × 1/2	76.07	HSS 355.6 × 254.0 × 12.7	113.2
HSS 14 × 10 × 3/8	58.10	HSS 355.6 × 254.0 × 9.5	86.3
HSS 14 × 10 × 5/16	48.86	HSS 355.6 × 254.0 × 7.9	72.4
HSS 14 × 10 × 1/4	39.43	HSS 355.6 × 254.0 × 6.4	59.1
HSS 14 × 6 × 5/8	76.33	HSS 355.6 × 152.4 × 15.9	113.8
HSS 14 × 6 × 1/2	62.46	HSS 355.6 × 152.4 × 12.7	93.0
HSS 14 × 6 × 3/8	47.90	HSS 355.6 × 152.4 × 9.5	71.1
HSS 14 × 6 × 5/16	40.35	HSS 355.6 × 152.4 × 7.9	59.8
HSS 14 × 6 × 1/4	32.63	HSS 355.6 × 152.4 × 6.4	48.9
HSS 14 × 6 × 3/16	24.73	HSS 355.6 × 152.4 × 4.8	37.1
HSS 14 × 4 × 5/8	67.82	HSS 355.6 × 101.6 × 15.9	101.1
HSS 14 × 4 × 1/2	55.66	HSS 355.6 × 101.6 × 12.7	82.8
HSS 14 × 4 × 3/8	42.79	HSS 355.6 × 101.6 × 9.5	63.5
HSS 14 × 4 × 5/16	36.1	HSS 355.6 × 101.6 × 7.9	53.5
HSS 14 × 4 × 1/4	29.23	HSS 355.6 × 101.6 × 6.4	43.8

STRUCTURAL STEEL WEIGHT TABLE (IMPERIAL AND METRIC) *(cont.)*

HSS (HOLLOW STRUCTURAL SECTIONS) RECTANGULAR

Imperial		Metric	
Designation	**Weight (lb/ft)**	**Designation**	**Weight (kg/m)**
HSS 14 × 4 × 3/16	22.18	HSS 355.6 × 101.6 × 4.8	33.3
HSS 12 × 10 × 1/2	69.27	HSS 304.8 × 254.0 × 12.7	103.1
HSS 12 × 10 × 3/8	53.00	HSS 304.8 × 254.0 × 9.5	78.7
HSS 12 × 10 × 5/16	44.60	HSS 304.8 × 254.0 × 7.9	66.1
HSS 12 × 10 × 1/4	36.03	HSS 304.8 × 254.0 × 6.4	54.0
HSS 12 × 8 × 5/8	76.33	HSS 304.8 × 203.2 × 15.9	113.8
HSS 12 × 8 × 1/2	62.46	HSS 304.8 × 203.2 × 12.7	93.0
HSS 12 × 8 × 3/8	47.90	HSS 304.8 × 203.2 × 9.5	71.1
HSS 12 × 8 × 5/16	40.35	HSS 304.8 × 203.2 × 7.9	59.8
HSS 12 × 8 × 1/4	32.63	HSS 304.8 × 203.2 × 6.4	48.9
HSS 12 × 8 × 3/16	24.73	HSS 304.8 × 203.2 × 4.8	37.1
HSS 12 × 6 × 5/8	67.82	HSS 304.8 × 152.4 × 15.9	101.1
HSS 12 × 6 × 1/2	55.66	HSS 304.8 × 152.4 × 12.7	82.8
HSS 12 × 6 × 3/8	42.79	HSS 304.8 × 152.4 × 9.5	63.5
HSS 12 × 6 × 5/16	36.10	HSS 304.8 × 152.4 × 7.9	53.5
HSS 12 × 6 × 1/4	29.23	HSS 304.8 × 152.4 × 6.4	43.8
HSS 12 × 6 × 3/16	22.18	HSS 304.8 × 152.4 × 4.8	33.3
HSS 12 × 4 × 5/8	59.32	HSS 304.8 × 101.6 × 15.9	88.4

STRUCTURAL STEEL WEIGHT TABLE
(IMPERIAL AND METRIC) *(cont.)*

HSS (HOLLOW STRUCTURAL SECTIONS) RECTANGULAR

Imperial		Metric	
Designation	Weight (lb/ft)	Designation	Weight (kg/m)
HSS 12 × 4 × 1/2	48.85	HSS 304.8 × 101.6 × 12.7	72.7
HSS 12 × 4 × 3/8	37.69	HSS 304.8 × 101.6 × 9.5	56.0
HSS 12 × 4 × 5/16	31.84	HSS 304.8 × 101.6 × 7.9	47.2
HSS 12 × 4 × 1/4	25.82	HSS 304.8 × 101.6 × 6.4	38.7
HSS 12 × 4 × 3/16	19.63	HSS 304.8 × 101.6 × 4.8	29.4
HSS 12 × 3-1/2 × 3/8	36.41	HSS 304.8 × 88.9 × 9.5	54.1
HSS 12 × 3-1/2 × 5/16	24.97	HSS 304.8 × 88.9 × 7.9	45.6
HSS 12 × 3 × 5/16	29.72	HSS 304.8 × 76.2 × 7.9	44.0
HSS 12 × 3 × 1/4	24.12	HSS 304.8 × 76.2 × 6.4	36.2
HSS 12 × 3 × 3/16	18.35	HSS 304.8 × 76.2 × 4.8	27.5
HSS 12 × 2 × 1/4	22.42	HSS 304.8 × 50.8 × 6.4	33.6
HSS 12 × 2 × 3/16	17.08	HSS 304.8 × 50.8 × 4.8	25.6
HSS 10 × 8 × 1/2	55.66	HSS 254.0 × 203.2 × 12.7	82.8
HSS 10 × 8 × 3/8	42.79	HSS 254.0 × 203.2 × 9.5	63.5
HSS 10 × 8 × 5/16	36.10	HSS 254.0 × 203.2 × 7.9	53.5
HSS 10 × 8 × 1/4	29.23	HSS 254.0 × 203.2 × 6.4	43.8
HSS 10 × 8 × 3/16	22.18	HSS 254.0 × 203.2 × 4.8	33.3
HSS 10 × 6 × 5/8	59.32	HSS 254.0 × 152.4 × 15.9	88.4

STRUCTURAL STEEL WEIGHT TABLE
(IMPERIAL AND METRIC) *(cont.)*

HSS (HOLLOW STRUCTURAL SECTIONS) RECTANGULAR

Imperial		Metric	
Designation	Weight (lb/ft)	Designation	Weight (kg/m)
HSS 10 × 6 × 1/2	48.85	HSS 254.0 × 152.4 × 12.7	72.7
HSS 10 × 6 × 3/8	37.69	HSS 254.0 × 152.4 × 9.5	56.0
HSS 10 × 6 × 5/16	31.84	HSS 254.0 × 152.4 × 7.9	47.2
HSS 10 × 6 × 1/4	25.82	HSS 254.0 × 152.4 × 6.4	38.7
HSS 10 × 6 × 3/16	19.63	HSS 254.0 × 152.4 × 4.8	29.4
HSS 10 × 5 × 3/8	35.13	HSS 254.0 × 127.0 × 9.5	52.2
HSS 10 × 5 × 5/16	29.72	HSS 254.0 × 127.0 × 7.9	44.0
HSS 10 × 5 × 1/4	24.12	HSS 254.0 × 127.0 × 6.4	36.2
HSS 10 × 5 × 3/16	18.35	HSS 254.0 × 127.0 × 4.8	27.5
HSS 10 × 4 × 5/8	50.81	HSS 254.0 × 101.6 × 15.9	75.7
HSS 10 × 4 × 1/2	42.05	HSS 254.0 × 101.6 × 12.7	62.6
HSS 10 × 4 × 3/8	32.58	HSS 254.0 × 101.6 × 9.5	48.4
HSS 10 × 4 × 5/16	27.59	HSS 254.0 × 101.6 × 7.9	40.9
HSS 10 × 4 × 1/4	22.42	HSS 254.0 × 101.6 × 6.4	33.6
HSS 10 × 4 × 3/16	17.08	HSS 254.0 × 101.6 × 4.8	25.6
HSS 10 × 3-1/2 × 3/16	16.44	HSS 254.0 × 88.9 × 4.8	24.7
HSS 10 × 3 × 3/8	30.03	HSS 254.0 × 76.2 × 9.5	44.6
HSS 10 × 3 × 5/16	25.46	HSS 254.0 × 76.2 × 7.9	37.7

STRUCTURAL STEEL WEIGHT TABLE (IMPERIAL AND METRIC) *(cont.)*

HSS (HOLLOW STRUCTURAL SECTIONS) RECTANGULAR

Imperial		Metric	
Designation	**Weight (lb/ft)**	**Designation**	**Weight (kg/m)**
HSS 10 × 3 × 1/4	20.72	HSS 254.0 × 76.2 × 6.4	31.1
HSS 10 × 3 × 3/16	15.80	HSS 254.0 × 76.2 × 4.8	23.7
HSS 10 × 3 × 1/8	10.71	HSS 254.0 × 76.2 × 3.2	16.1
HSS 10 × 2 × 3/8	27.48	HSS 254.0 × 50.8 × 9.5	40.8
HSS 10 × 2 × 5/16	23.34	HSS 254.0 × 50.8 × 7.9	34.6
HSS 10 × 2 × 1/4	19.02	HSS 254.0 × 50.8 × 6.4	28.5
HSS 10 × 2 × 3/16	14.53	HSS 254.0 × 50.8 × 4.8	21.8
HSS 9 × 7 × 5/8	59.32	HSS 228.6 × 177.8 × 15.9	88.4
HSS 9 × 7 × 1/2	48.85	HSS 228.6 × 177.8 × 12.7	72.7
HSS 9 × 7 × 3/8	37.69	HSS 228.6 × 177.8 × 9.5	56.0
HSS 9 × 7 × 5/16	31.84	HSS 228.6 × 177.8 × 7.9	47.2
HSS 9 × 7 × 1/4	25.82	HSS 228.6 × 177.8 × 6.4	38.7
HSS 9 × 7 × 3/16	19.63	HSS 228.6 × 177.8 × 4.8	29.4
HSS 9 × 5 × 5/8	50.81	HSS 228.6 × 127.0 × 15.9	75.7
HSS 9 × 5 × 1/2	42.05	HSS 228.6 × 127.0 × 12.7	62.6
HSS 9 × 5 × 3/8	32.58	HSS 228.6 × 127.0 × 9.5	48.4
HSS 9 × 5 × 5/16	27.59	HSS 228.6 × 127.0 × 7.9	40.9
HSS 9 × 5 × 1/4	22.42	HSS 228.6 × 127.0 × 6.4	33.6

STRUCTURAL STEEL WEIGHT TABLE (IMPERIAL AND METRIC) *(cont.)*

HSS (HOLLOW STRUCTURAL SECTIONS) RECTANGULAR

Imperial		Metric	
Designation	**Weight (lb/ft)**	**Designation**	**Weight (kg/m)**
HSS 9 × 5 × 3/16	17.08	HSS 228.6 × 127.0 × 4.8	25.6
HSS 9 × 3 × 1/2	35.24	HSS 228.6 × 76.2 × 12.7	52.4
HSS 9 × 3 × 3/8	27.48	HSS 228.6 × 76.2 × 9.5	40.8
HSS 9 × 3 × 5/16	23.34	HSS 228.6 × 76.2 × 7.9	34.6
HSS 9 × 3 × 1/4	19.02	HSS 228.6 × 76.2 × 6.4	28.5
HSS 9 × 3 × 3/16	14.53	HSS 228.6 × 76.2 × 4.8	21.8
HSS 8 × 6 × 5/8	50.81	HSS 203.2 × 152.4 × 15.9	75.7
HSS 8 × 6 × 1/2	42.05	HSS 203.2 × 152.4 × 12.7	62.6
HSS 8 × 6 × 3/8	32.58	HSS 203.2 × 152.4 × 9.5	48.4
HSS 8 × 6 × 5/16	27.59	HSS 203.2 × 152.4 × 7.9	40.9
HSS 8 × 6 × 1/4	22.42	HSS 203.2 × 152.4 × 6.4	33.6
HSS 8 × 6 × 3/16	17.08	HSS 203.2 × 152.4 × 4.8	25.6
HSS 8 × 4 × 5/8	42.30	HSS 203.2 × 101.6 × 15.9	63.0
HSS 8 × 4 × 1/2	35.24	HSS 203.2 × 101.6 × 12.7	52.4
HSS 8 × 4 × 3/8	27.48	HSS 203.2 × 101.6 × 9.5	40.8
HSS 8 × 4 × 5/16	23.34	HSS 203.2 × 101.6 × 7.9	34.6
HSS 8 × 4 × 1/4	19.02	HSS 203.2 × 101.6 × 6.4	28.5
HSS 8 × 4 × 3/16	14.53	HSS 203.2 × 101.6 × 4.8	21.8

STRUCTURAL STEEL WEIGHT TABLE
(IMPERIAL AND METRIC) *(cont.)*

HSS (HOLLOW STRUCTURAL SECTIONS) RECTANGULAR

Imperial		Metric	
Designation	Weight (lb/ft)	Designation	Weight (kg/m)
HSS 8 × 4 × 1/8	9.86	HSS 203.2 × 101.6 × 3.2	14.8
HSS 8 × 3 × 1/2	31.84	HSS 203.2 × 76.2 × 12.7	47.4
HSS 8 × 3 × 3/8	24.93	HSS 203.2 × 76.2 × 9.5	37.0
HSS 8 × 3 × 5/16	21.21	HSS 203.2 × 76.2 × 7.9	31.4
HSS 8 × 3 × 1/4	17.32	HSS 203.2 × 76.2 × 6.4	26.0
HSS 8 × 3 × 3/16	13.25	HSS 203.2 × 76.2 × 4.8	19.9
HSS 8 × 3 × 1/8	9.01	HSS 203.2 × 76.2 × 3.2	13.5
HSS 8 × 2 × 3/8	22.37	HSS 203.2 × 50.8 × 9.5	33.2
HSS 8 × 2 × 5/16	19.08	HSS 203.2 × 50.8 × 7.9	28.3
HSS 8 × 2 × 1/4	15.62	HSS 203.2 × 50.8 × 6.4	23.4
HSS 8 × 2 × 3/16	11.97	HSS 203.2 × 50.8 × 4.8	18.0
HSS 8 × 2 × 1/8	8.16	HSS 203.2 × 50.8 × 3.2	12.2
HSS 7 × 5 × 5/8	42.30	HSS 177.8 × 127.0 × 15.9	63.0
HSS 7 × 5 × 1/2	35.24	HSS 177.8 × 127.0 × 12.7	52.4
HSS 7 × 5 × 3/8	27.48	HSS 177.8 × 127.0 × 9.5	40.8
HSS 7 × 5 × 5/16	23.34	HSS 177.8 × 127.0 × 7.9	34.6
HSS 7 × 5 × 1/4	19.02	HSS 177.8 × 127.0 × 6.4	28.5
HSS 7 × 5 × 3/16	14.53	HSS 177.8 × 127.0 × 4.8	21.8

STRUCTURAL STEEL WEIGHT TABLE (IMPERIAL AND METRIC) (cont.)

HSS (HOLLOW STRUCTURAL SECTIONS) RECTANGULAR

Imperial		Metric	
Designation	Weight (lb/ft)	Designation	Weight (kg/m)
HSS 7 × 5 × 1/8	9.86	HSS 177.8 × 127.0 × 3.2	14.8
HSS 7 × 4 × 1/2	31.84	HSS 177.8 × 101.6 × 12.7	47.4
HSS 7 × 4 × 3/8	24.93	HSS 177.8 × 101.6 × 9.5	37.0
HSS 7 × 4 × 5/16	21.21	HSS 177.8 × 101.6 × 7.9	31.4
HSS 7 × 4 × 1/4	17.32	HSS 177.8 × 101.6 × 6.4	26.0
HSS 7 × 4 × 3/16	13.25	HSS 177.8 × 101.6 × 4.8	19.9
HSS 7 × 4 × 1/8	9.01	HSS 177.8 × 101.6 × 3.2	13.5
HSS 7 × 3 × 1/2	28.43	HSS 177.8 × 76.2 × 12.7	42.3
HSS 7 × 3 × 3/8	22.37	HSS 177.8 × 76.2 × 9.5	33.2
HSS 7 × 3 × 5/16	19.08	HSS 177.8 × 76.2 × 7.9	28.3
HSS 7 × 3 × 1/4	15.62	HSS 177.8 × 76.2 × 6.4	23.4
HSS 7 × 3 × 3/16	11.97	HSS 177.8 × 76.2 × 4.8	18.0
HSS 7 × 3 × 1/8	8.16	HSS 177.8 × 76.2 × 3.2	12.2
HSS 6 × 5 × 3/8	24.93	HSS 152.4 × 127.0 × 9.5	37.0
HSS 6 × 5 × 5/16	21.21	HSS 152.4 × 127.0 × 7.9	31.4
HSS 6 × 5 × 1/4	17.32	HSS 152.4 × 127.0 × 6.4	26.0
HSS 6 × 5 × 3/16	13.25	HSS 152.4 × 127.0 × 4.8	19.9
HSS 6 × 4 × 1/2	28.43	HSS 152.4 × 101.6 × 12.7	42.3

STRUCTURAL STEEL WEIGHT TABLE (IMPERIAL AND METRIC) *(cont.)*

HSS (HOLLOW STRUCTURAL SECTIONS) RECTANGULAR

Imperial		Metric	
Designation	Weight (lb/ft)	Designation	Weight (kg/m)
HSS 6 × 4 × 3/8	22.37	HSS 152.4 × 101.6 × 9.5	33.2
HSS 6 × 4 × 5/16	19.08	HSS 152.4 × 101.6 × 7.9	28.3
HSS 6 × 4 × 1/4	15.62	HSS 152.4 × 101.6 × 6.4	23.4
HSS 6 × 4 × 3/16	11.97	HSS 152.4 × 101.6 × 4.8	18.0
HSS 6 × 4 × 1/8	8.16	HSS 152.4 × 101.6 × 3.2	12.2
HSS 6 × 3 × 1/2	25.03	HSS 152.4 × 76.2 × 12.7	37.3
HSS 6 × 3 × 3/8	19.82	HSS 152.4 × 76.2 × 9.5	29.4
HSS 6 × 3 × 5/16	16.96	HSS 152.4 × 76.2 × 7.9	25.1
HSS 6 × 3 × 1/4	13.91	HSS 152.4 × 76.2 × 6.4	20.9
HSS 6 × 3 × 3/16	10.70	HSS 152.4 × 76.2 × 4.8	16.0
HSS 6 × 3 × 1/8	7.31	HSS 152.4 × 76.2 × 3.2	11.0
HSS 6 × 2 × 3/8	17.27	HSS 152.4 × 50.8 × 9.5	25.6
HSS 6 × 2 × 5/16	14.83	HSS 152.4 × 50.8 × 7.9	22.0
HSS 6 × 2 × 1/4	12.21	HSS 152.4 × 50.8 × 6.4	18.3
HSS 6 × 2 × 3/16	9.42	HSS 152.4 × 50.8 × 4.8	14.1
HSS 6 × 2 × 1/8	6.46	HSS 152.4 × 50.8 × 3.2	9.7
HSS 5 × 4 × 1/2	25.03	HSS 127.0 × 101.6 × 12.7	37.3
HSS 5 × 4 × 3/8	19.82	HSS 127.0 × 101.6 × 9.5	29.4

STRUCTURAL STEEL WEIGHT TABLE (IMPERIAL AND METRIC) *(cont.)*

HSS (HOLLOW STRUCTURAL SECTIONS) RECTANGULAR

Imperial		Metric	
Designation	**Weight (lb/ft)**	**Designation**	**Weight (kg/m)**
HSS 5 × 4 × 5/16	16.96	HSS 127.0 × 101.6 × 7.9	25.1
HSS 5 × 4 × 1/4	13.91	HSS 127.0 × 101.6 × 6.4	20.9
HSS 5 × 4 × 3/16	10.70	HSS 127.0 × 101.6 × 4.8	16.0
HSS 5 × 3 × 1/2	21.63	HSS 127.0 × 76.2 × 12.7	32.2
HSS 5 × 3 × 3/8	17.27	HSS 127.0 × 76.2 × 9.5	25.6
HSS 5 × 3 × 5/16	14.83	HSS 127.0 × 76.2 × 7.9	22.0
HSS 5 × 3 × 1/4	12.21	HSS 127.0 × 76.2 × 6.4	18.3
HSS 5 × 3 × 3/16	9.42	HSS 127.0 × 76.2 × 4.8	14.1
HSS 5 × 3 × 1/8	6.46	HSS 127.0 × 76.2 × 3.2	9.7
HSS 5 × 2-1/2 × 1/4	11.36	HSS 127.0 × 63.5 × 6.4	17.0
HSS 5 × 2-1/2 × 3/16	8.78	HSS 127.0 × 63.5 × 4.8	13.2
HSS 5 × 2-1/2 × 1/8	6.03	HSS 127.0 × 63.5 × 3.2	9.0
HSS 5 × 2 × 3/8	14.72	HSS 127.0 × 50.8 × 9.5	21.9
HSS 5 × 2 × 5/16	12.70	HSS 127.0 × 50.8 × 7.9	18.8
HSS 5 × 2 × 1/4	10.51	HSS 127.0 × 50.8 × 6.4	15.8
HSS 5 × 2 × 3/16	8.15	HSS 127.0 × 50.8 × 4.8	12.2
HSS 5 × 2 × 1/8	5.61	HSS 127.0 × 50.8 × 3.2	8.4
HSS 4 × 3 × 3/8	14.72	HSS 101.6 × 76.2 × 9.5	21.9

STRUCTURAL STEEL WEIGHT TABLE
(IMPERIAL AND METRIC) *(cont.)*

HSS (HOLLOW STRUCTURAL SECTIONS) RECTANGULAR

Imperial		Metric	
Designation	**Weight (lb/ft)**	**Designation**	**Weight (kg/m)**
HSS 4 × 3 × 5/16	12.70	HSS 127.0 × 50.8 × 7.9	18.8
HSS 4 × 3 × 1/4	10.51	HSS 127.0 × 50.8 × 6.4	15.8
HSS 4 × 3 × 3/16	8.15	HSS 127.0 × 50.8 × 4.8	12.2
HSS 4 × 3 × 1/8	5.61	HSS 127.0 × 50.8 × 3.2	8.4
HSS 4 × 2-1/2 × 5/16	11.64	HSS 101.6 × 63.5 × 7.9	17.3
HSS 4 × 2-1/2 × 1/4	9.66	HSS 101.6 × 63.5 × 6.4	14.5
HSS 4 × 2-1/2 × 3/16	7.51	HSS 101.6 × 63.5 × 4.8	11.3
HSS 4 × 2 × 3/8	12.17	HSS 101.6 × 50.8 × 9.5	18.1
HSS 4 × 2 × 5/16	10.58	HSS 101.6 × 50.8 × 7.9	15.7
HSS 4 × 2 × 1/4	8.81	HSS 101.6 × 50.8 × 6.4	13.2
HSS 4 × 2 × 3/16	6.87	HSS 101.6 × 50.8 × 4.8	10.3
HSS 4 × 2 × 1/8	4.75	HSS 101.6 × 50.8 × 3.2	7.1
HSS 3-1/2 × 2-1/2 × 3/8	12.17	HSS 88.9 × 63.5 × 9.5	18.1
HSS 3-1/2 × 2-1/2 × 5/16	10.58	HSS 101.6 × 50.8 × 7.9	15.7
HSS 3-1/2 × 2-1/2 × 1/4	8.81	HSS 101.6 × 50.8 × 6.4	13.2
HSS 3-1/2 × 2-1/2 × 3/16	6.87	HSS 101.6 × 50.8 × 4.8	10.3
HSS 3-1/2 × 2-1/2 × 1/8	4.75	HSS 101.6 × 50.8 × 3.2	7.1
HSS 3 × 2-1/2 × 5/16	9.51	HSS 76.2 × 63.5 × 7.9	14.1

STRUCTURAL STEEL WEIGHT TABLE
(IMPERIAL AND METRIC) *(cont.)*

HSS (HOLLOW STRUCTURAL SECTIONS) RECTANGULAR

Imperial		Metric	
Designation	**Weight (lb/ft)**	**Designation**	**Weight (kg/m)**
HSS 3 × 2-1/2 × 1/4	7.96	HSS 76.2 × 63.5 × 6.4	11.9
HSS 3 × 2-1/2 × 3/16	6.23	HSS 76.2 × 63.5 × 4.8	9.3
HSS 3 × 2-1/2 × 1/8	4.33	HSS 76.2 × 63.5 × 3.2	6.5
HSS 3 × 2 × 5/16	8.45	HSS 76.2 × 50.8 × 7.9	12.5
HSS 3 × 2 × 1/4	7.11	HSS 76.2 × 50.8 × 6.4	10.6
HSS 3 × 2 × 3/16	5.59	HSS 76.2 × 50.8 × 4.8	8.4
HSS 3 × 2 × 1/8	3.90	HSS 76.2 × 50.8 × 3.2	5.9
HSS 3 × 1-1/2 × 1/4	6.26	HSS 76.2 × 38.1 × 6.4	9.4
HSS 3 × 1-1/2 × 3/16	4.96	HSS 76.2 × 38.1 × 4.8	7.4
HSS 3 × 1-1/2 × 1/8	3.48	HSS 76.2 × 38.1 × 3.2	5.2
HSS 3 × 1 × 3/16	6.32	HSS 76.2 × 25.4 × 4.8	6.5
HSS 3 × 1 × 1/8	3.05	HSS 76.2 × 38.1 × 3.2	4.6
HSS 2-1/2 × 1-1/2 × 1/4	5.41	HSS 63.5 × 38.1 × 6.4	8.1
HSS 2-1/2 × 1-1/2 × 3/16	4.32	HSS 63.5 × 38.1 × 4.8	6.5
HSS 2-1/2 × 1-1/2 × 1/8	3.05	HSS 63.5 × 38.1 × 3.2	4.6
HSS 2 × 1-1/2 × 3/16	3.68	HSS 50.8 × 38.1 × 4.8	5.5
HSS 2 × 1-1/2 × 1/8	2.63	HSS 50.8 × 38.1 × 3.2	3.9
HSS 2 × 1 × 3/16	3.04	HSS 50.8 × 25.4 × 4.8	4.6
HSS 2 × 1 × 1/8	2.20	HSS 50.8 × 25.4 × 3.2	3.3

STRUCTURAL STEEL WEIGHT TABLE (IMPERIAL AND METRIC) *(cont.)*

HSS (HOLLOW STRUCTURAL SECTIONS) SQUARE

| Imperial | | Metric | |
Designation	Weight (lb/ft)	Designation	Weight (kg/m)
HSS 32 × 32 × 5/8	259.83	HSS 812.8 × 812.8 × 15.9	387.3
HSS 32 × 32 × 1/2	210.72	HSS 812.8 × 812.8 × 12.7	313.6
HSS 32 × 32 × 3/8	159.37	HSS 812.8 × 812.8 × 9.5	236.6
HSS 30 × 30 × 5/8	242.82	HSS 762.0 × 762.0 × 15.9	361.9
HSS 30 × 30 × 1/2	197.11	HSS 762.0 × 762.0 × 12.7	293.4
HSS 30 × 30 × 3/8	149.16	HSS 762.0 × 762.0 × 9.5	221.4
HSS 28 × 28 × 5/8	225.80	HSS 711.2 × 711.2 × 15.9	336.6
HSS 28 × 28 × 1/2	183.50	HSS 711.2 × 711.2 × 12.7	273.1
HSS 28 × 28 × 3/8	138.95	HSS 711.2 × 711.2 × 9.5	206.3
HSS 26 × 26 × 5/8	208.79	HSS 660.4 × 660.4 × 15.9	311.2
HSS 26 × 26 × 1/2	169.89	HSS 660.4 × 660.4 × 12.7	252.9
HSS 26 × 26 × 3/8	128.74	HSS 660.4 × 660.4 × 9.5	191.1
HSS 24 × 24 × 5/8	191.78	HSS 609.6 × 609.6 × 15.9	285.8
HSS 24 × 24 × 1/2	156.28	HSS 609.6 × 609.6 × 12.7	232.6
HSS 24 × 24 × 3/8	118.53	HSS 609.6 × 609.6 × 9.5	176.0
HSS 22 × 22 × 5/8	174.76	HSS 558.8 × 558.8 × 15.9	260.5
HSS 22 × 22 × 1/2	142.67	HSS 558.8 × 558.8 × 12.7	212.3
HSS 22 × 22 × 3/8	108.32	HSS 558.8 × 558.8 × 9.5	160.8
HSS 20 × 20 × 5/8	157.75	HSS 508.0 × 508.0 × 15.9	235.1
HSS 20 × 20 × 1/2	129.06	HSS 508.0 × 508.0 × 12.7	192.1

STRUCTURAL STEEL WEIGHT TABLE (IMPERIAL AND METRIC) *(cont.)*

HSS (HOLLOW STRUCTURAL SECTIONS) SQUARE

Imperial		Metric	
Designation	**Weight (lb/ft)**	**Designation**	**Weight (kg/m)**
HSS 20 × 20 × 3/8	98.12	HSS 508.0 × 508.0 × 9.5	145.7
HSS 18 × 18 × 5/8	140.73	HSS 457.2 × 457.2 × 15.9	209.8
HSS 18 × 18 × 1/2	115.45	HSS 457.2 × 457.2 × 12.7	171.8
HSS 18 × 18 × 3/8	87.91	HSS 457.2 × 457.2 × 9.5	130.5
HSS 16 × 16 × 5/8	127.37	HSS 406.4 × 406.4 × 15.9	189.9
HSS 16 × 16 × 1/2	103.30	HSS 406.4 × 406.4 × 12.7	153.7
HSS 16 × 16 × 3/8	78.52	HSS 406.4 × 406.4 × 9.5	116.6
HSS 16 × 16 × 5/16	65.87	HSS 406.4 × 406.4 × 7.9	97.6
HSS 14 × 14 × 5/8	110.36	HSS 355.6 × 355.6 × 15.9	164.5
HSS 14 × 14 × 1/2	89.68	HSS 355.6 × 355.6 × 12.7	133.5
HSS 14 × 14 × 3/8	68.31	HSS 355.6 × 355.6 × 9.5	101.4
HSS 14 × 14 × 5/16	57.36	HSS 355.6 × 355.6 × 7.9	85.0
HSS 12 × 12 × 5/8	93.34	HSS 304.8 × 304.8 × 15.9	139.1
HSS 12 × 12 × 1/2	76.07	HSS 304.8 × 304.8 × 12.7	113.2
HSS 12 × 12 × 3/8	58.10	HSS 304.8 × 304.8 × 9.5	86.3
HSS 12 × 12 × 5/16	48.86	HSS 304.8 × 304.8 × 7.9	72.4
HSS 12 × 12 × 1/4	39.43	HSS 304.8 × 304.8 × 6.4	59.1
HSS 10 × 10 × 5/8	76.33	HSS 254.0 × 254.0 × 15.9	113.8
HSS 10 × 10 × 1/2	62.46	HSS 254.0 × 254.0 × 12.7	93.0
HSS 10 × 10 × 3/8	47.90	HSS 254.0 × 254.0 × 9.5	71.1

STRUCTURAL STEEL WEIGHT TABLE
(IMPERIAL AND METRIC) *(cont.)*

HSS (HOLLOW STRUCTURAL SECTIONS) SQUARE

Imperial		Metric	
Designation	**Weight (lb/ft)**	**Designation**	**Weight (kg/m)**
HSS 10 × 10 × 5/16	40.35	HSS 254.0 × 254.0 × 7.9	59.8
HSS 10 × 10 × 1/4	32.63	HSS 254.0 × 254.0 × 6.4	48.9
HSS 10 × 10 × 3/16	24.73	HSS 254.0 × 254.0 × 4.8	37.1
HSS 9 × 9 × 1/2	55.66	HSS 228.6 × 228.6 × 12.7	82.8
HSS 9 × 9 × 3/8	42.79	HSS 228.6 × 228.6 × 9.5	63.5
HSS 9 × 9 × 5/16	36.1	HSS 228.6 × 228.6 × 7.9	53.5
HSS 9 × 9 × 1/4	29.23	HSS 228.6 × 228.6 × 6.4	43.8
HSS 9 × 9 × 3/16	22.18	HSS 228.6 × 228.6 × 4.8	33.3
HSS 8 × 8 × 5/8	59.32	HSS 203.2 × 203.2 × 15.9	88.4
HSS 8 × 8 × 1/2	48.85	HSS 203.2 × 203.2 × 12.7	72.7
HSS 8 × 8 × 3/8	37.69	HSS 203.2 × 203.2 × 9.5	56.0
HSS 8 × 8 × 5/16	31.84	HSS 203.2 × 203.2 × 7.9	47.2
HSS 8 × 8 × 1/4	25.82	HSS 203.2 × 203.2 × 6.4	38.7
HSS 8 × 8 × 3/16	19.63	HSS 203.2 × 203.2 × 4.8	29.4
HSS 7 × 7 × 5/8	50.81	HSS 177.8 × 177.8 × 15.9	75.7
HSS 7 × 7 × 1/2	42.05	HSS 177.8 × 177.8 × 12.7	62.6
HSS 7 × 7 × 3/8	32.58	HSS 177.8 × 177.8 × 9.5	48.4
HSS 7 × 7 × 5/16	27.59	HSS 177.8 × 177.8 × 7.9	40.9
HSS 7 × 7 × 1/4	22.42	HSS 177.8 × 177.8 × 6.4	33.6
HSS 7 × 7 × 3/16	17.08	HSS 177.8 × 177.8 × 4.8	25.6

STRUCTURAL STEEL WEIGHT TABLE
(IMPERIAL AND METRIC) *(cont.)*

HSS (HOLLOW STRUCTURAL SECTIONS) SQUARE

Imperial		Metric	
Designation	Weight (lb/ft)	Designation	Weight (kg/m)
HSS 6 × 6 × 5/8	42.30	HSS 152.4 × 152.4 × 15.9	63.0
HSS 6 × 6 × 1/2	35.24	HSS 152.4 × 152.4 × 12.7	52.4
HSS 6 × 6 × 3/8	27.48	HSS 152.4 × 152.4 × 9.5	40.8
HSS 6 × 6 × 5/16	23.34	HSS 152.4 × 152.4 × 7.9	34.6
HSS 6 × 6 × 1/4	19.02	HSS 152.4 × 152.4 × 6.4	28.5
HSS 6 × 6 × 3/16	14.53	HSS 152.4 × 152.4 × 4.8	21.8
HSS 6 × 6 × 1/8	9.86	HSS 152.4 × 152.4 × 3.2	14.8
HSS 5-1/2 × 5-1/2 × 3/8	24.93	HSS 139.7 × 139.7 × 9.5	37.0
HSS 5-1/2 × 5-1/2 × 5/16	21.21	HSS 139.7 × 139.7 × 7.9	31.4
HSS 5-1/2 × 5-1/2 × 1/4	17.32	HSS 139.7 × 139.7 × 6.4	26.0
HSS 5-1/2 × 5-1/2 × 3/16	13.25	HSS 139.7 × 139.7 × 4.8	19.9
HSS 5-1/2 × 5-1/2 × 1/8	9.01	HSS 139.7 × 139.7 × 3.2	13.5
HSS 5 × 5 × 1/2	28.43	HSS 127.0 × 127.0 × 12.7	42.3
HSS 5 × 5 × 3/8	22.37	HSS 127.0 × 127.0 × 9.5	33.2
HSS 5 × 5 × 5/16	19.08	HSS 127.0 × 127.0 × 7.9	28.3
HSS 5 × 5 × 1/4	15.62	HSS 127.0 × 127.0 × 6.4	23.4
HSS 5 × 5 × 3/16	11.97	HSS 127.0 × 127.0 × 4.8	18.0
HSS 5 × 5 × 1/8	8.16	HSS 127.0 × 127.0 × 3.2	12.2
HSS 4-1/2 × 4-1/2 × 1/2	25.03	HSS 114.3 × 114.3 × 12.7	37.3
HSS 4-1/2 × 4-1/2 × 3/8	19.82	HSS 114.3 × 114.3 × 9.5	29.4

STRUCTURAL STEEL WEIGHT TABLE (IMPERIAL AND METRIC) *(cont.)*

HSS (HOLLOW STRUCTURAL SECTIONS) SQUARE

Imperial		Metric	
Designation	Weight (lb/ft)	Designation	Weight (kg/m)
HSS 4-1/2 × 4-1/2 × 5/16	16.96	HSS 114.3 × 114.3 × 7.9	25.1
HSS 4-1/2 × 4-1/2 × 1/4	13.91	HSS 114.3 × 114.3 × 6.4	20.9
HSS 4-1/2 × 4-1/2 × 3/16	10.7	HSS 114.3 × 114.3 × 4.8	16.0
HSS 4-1/2 × 4-1/2 × 1/8	7.31	HSS 114.3 × 114.3 × 3.2	11.0
HSS 4 × 4 × 1/2	21.63	HSS 101.6 × 101.6 × 12.7	32.2
HSS 4 × 4 × 3/8	17.27	HSS 101.6 × 101.6 × 9.5	25.6
HSS 4 × 4 × 5/16	14.83	HSS 101.6 × 101.6 × 7.9	22.0
HSS 4 × 4 × 1/4	12.21	HSS 101.6 × 101.6 × 6.4	18.3
HSS 4 × 4 × 3/16	9.42	HSS 101.6 × 101.6 × 4.8	14.1
HSS 4 × 4 × 1/8	6.46	HSS 101.6 × 101.6 × 3.2	9.7
HSS 3-1/2 × 3-1/2 × 3/8	14.72	HSS 88.9 × 88.9 × 9.5	21.9
HSS 3-1/2 × 3-1/2 × 5/16	12.70	HSS 88.9 × 88.9 × 7.9	18.8
HSS 3-1/2 × 3-1/2 × 1/4	10.51	HSS 88.9 × 88.9 × 6.4	15.8
HSS 3-1/2 × 3-1/2 × 3/16	8.15	HSS 88.9 × 88.9 × 4.8	12.2
HSS 3-1/2 × 3-1/2 × 1/8	5.61	HSS 88.9 × 88.9 × 3.2	8.4
HSS 3 × 3 × 3/8	12.17	HSS 76.2 × 76.2 × 9.5	18.1
HSS 3 × 3 × 5/16	10.58	HSS 76.2 × 76.2 × 7.9	15.7
HSS 3 × 3 × 1/4	8.81	HSS 76.2 × 76.2 × 6.4	13.2
HSS 3 × 3 × 3/16	6.87	HSS 76.2 × 76.2 × 4.8	10.3
HSS 3 × 3 × 1/8	4.75	HSS 76.2 × 76.2 × 3.2	7.1

STRUCTURAL STEEL WEIGHT TABLE
(IMPERIAL AND METRIC) *(cont.)*

HSS (HOLLOW STRUCTURAL SECTIONS) SQUARE

Imperial		Metric	
Designation	Weight (lb/ft)	Designation	Weight (kg/m)
HSS 2-1/2 × 2-1/2 × 5/16	8.45	HSS 63.5 × 63.5 × 7.9	12.5
HSS 2-1/2 × 2-1/2 × 1/4	7.11	HSS 63.5 × 63.5 × 6.4	10.6
HSS 2-1/2 × 2-1/2 × 3/16	5.59	HSS 63.5 × 63.5 × 4.8	8.4
HSS 2-1/2 × 2-1/2 × 1/8	3.90	HSS 63.5 × 63.5 × 3.2	5.9
HSS 2-1/4 × 2-1/4 × 1/4	6.26	HSS 57.2 × 57.2 × 6.4	9.4
HSS 2-1/4 × 2-1/4 × 3/16	4.96	HSS 57.2 × 57.2 × 4.8	7.4
HSS 2-1/4 × 2-1/4 × 1/8	3.48	HSS 57.2 × 57.2 × 3.2	5.2
HSS 2 × 2 × 1/4	5.41	HSS 50.8 × 50.8 × 6.4	8.1
HSS 2 × 2 × 3/16	4.32	HSS 50.8 × 50.8 × 4.8	6.5
HSS 2 × 2 × 1/8	3.05	HSS 50.8 × 50.8 × 3.2	4.6
HSS 1-3/4 × 1-3/4 × 3/16	3.68	HSS 44.5 × 44.5 × 4.8	5.5
HSS 1-5/8 × 1-5/8 × 3/16	3.36	HSS 41.3 × 41.3 × 4.8	5.0
HSS 1-5/8 × 1-5/8 × 1/8	2.42	HSS 41.3 × 41.3 × 3.2	3.6
HSS 1-1/2 × 1-1/2 × 3/16	3.04	HSS 38.1 × 38.1 × 4.8	4.6
HSS 1-1/2 × 1-1/2 × 1/8	2.20	HSS 38.1 × 38.1 × 3.2	3.3
HSS 1-1/4 × 1-1/4 × 3/16	2.40	HSS 31.8 × 31.8 × 4.8	3.6
HSS 1-1/4 × 1-1/4 × 1/8	1.78	HSS 31.8 × 31.8 × 3.2	2.7

STRUCTURAL STEEL WEIGHT TABLE
(IMPERIAL AND METRIC) *(cont.)*

HSS (HOLLOW STRUCTURAL SECTIONS) ROUND

Imperial		Metric	
Designation	**Weight (lb/ft)**	**Designation**	**Weight (kg/m)**
HSS 20.000 × 0.500	104.23	HSS 508.0 × 12.7	155.1
HSS 20.000 × 0.375	78.67	HSS 508.0 × 9.5	116.8
HSS 18.000 × 0.500	93.54	HSS 457.2 × 12.7	139.2
HSS 18.000 × 0.375	70.65	HSS 457.2 × 9.5	104.9
HSS 16.000 × 0.500	82.85	HSS 406.4 × 12.7	123.3
HSS 16.000 × 0.438	72.86	HSS 406.4 × 11.1	108.2
HSS 16.000 × 0.375	62.64	HSS 406.4 × 9.5	93.0
HSS 16.000 × 0.312	52.32	HSS 406.4 × 7.9	77.6
HSS 14.000 × 0.500	72.16	HSS 355.6 × 12.7	107.4
HSS 14.000 × 0.375	54.62	HSS 355.6 × 9.5	81.1
HSS 14.000 × 0.312	45.65	HSS 355.6 × 7.9	67.7
HSS 12.750 × 0.500	65.48	HSS 323.9 × 12.7	97.5
HSS 12.750 × 0.375	49.61	HSS 323.9 × 9.5	73.7
HSS 12.750 × 0.250	33.41	HSS 323.9 × 6.4	50.1
HSS 12.500 × 0.625	79.34	HSS 317.5 × 15.9	118.3
HSS 12.500 × 0.500	64.14	HSS 317.5 × 12.7	95.5
HSS 12.500 × 0.375	48.61	HSS 317.5 × 9.5	72.2
HSS 12.500 × 0.312	40.65	HSS 317.5 × 7.9	60.3
HSS 12.500 × 0.250	32.74	HSS 317.5 × 6.4	49.1
HSS 12.500 × 0.188	24.74	HSS 317.5 × 4.8	37.0

STRUCTURAL STEEL WEIGHT TABLE (IMPERIAL AND METRIC) *(cont.)*

HSS (HOLLOW STRUCTURAL SECTIONS) ROUND

Imperial		Metric	
Designation	Weight (lb/ft)	Designation	Weight (kg/m)
HSS 12.313 × 0.625	78.09	HSS 312.8 × 15.9	116.4
HSS 12.313 × 0.500	63.14	HSS 312.8 × 12.7	94.0
HSS 12.313 × 0.375	47.86	HSS 312.8 × 9.5	71.0
HSS 12.313 × 0.312	40.03	HSS 312.8 × 7.9	59.4
HSS 12.313 × 0.250	32.24	HSS 312.8 × 6.4	48.4
HSS 12.313 × 0.188	24.37	HSS 312.8 × 4.8	36.5
HSS 12.250 × 0.625	77.67	HSS 311.2 × 15.9	115.8
HSS 12.250 × 0.500	62.80	HSS 311.2 × 12.7	93.5
HSS 12.250 × 0.375	47.60	HSS 311.2 × 9.5	70.7
HSS 12.250 × 0.312	39.82	HSS 311.2 × 7.9	59.1
HSS 12.250 × 0.250	32.07	HSS 311.2 × 6.4	48.1
HSS 12.250 × 0.188	24.24	HSS 311.2 × 4.8	36.3
HSS 11.250 × 0.625	70.99	HSS 285.8 × 15.9	105.8
HSS 11.250 × 0.500	57.46	HSS 285.8 × 12.7	85.5
HSS 11.250 × 0.375	43.60	HSS 285.8 × 9.5	64.7
HSS 11.250 × 0.312	36.48	HSS 285.8 × 7.9	54.1
HSS 11.250 × 0.250	29.40	HSS 285.8 × 6.4	44.1
HSS 11.250 × 0.188	22.23	HSS 285.8 × 4.8	33.3
HSS 10.750 × 0.500	54.79	HSS 273.1 × 12.7	81.6
HSS 10.750 × 0.365	40.52	HSS 273.1 × 9.3	60.5

STRUCTURAL STEEL WEIGHT TABLE
(IMPERIAL AND METRIC) *(cont.)*

HSS (HOLLOW STRUCTURAL SECTIONS) ROUND

Imperial		Metric	
Designation	Weight (lb/ft)	Designation	Weight (kg/m)
HSS 10.750 × 0.250	28.06	HSS 273.1 × 6.4	42.1
HSS 10.000 × 0.625	62.64	HSS 254.0 × 15.9	93.4
HSS 10.000 × 0.500	50.78	HSS 254.0 × 12.7	75.6
HSS 10.000 × 0.375	38.58	HSS 254.0 × 9.5	57.3
HSS 10.000 × 0.312	32.31	HSS 254.0 × 7.9	47.9
HSS 10.000 × 0.250	26.06	HSS 254.0 × 6.4	39.1
HSS 10.000 × 0.188	19.72	HSS 254.0 × 4.8	29.5
HSS 9.625 × 0.500	48.77	HSS 244.5 × 12.7	72.6
HSS 9.625 × 0.375	37.08	HSS 244.5 × 9.5	55.1
HSS 9.625 × 0.312	31.06	HSS 244.5 × 7.9	46.1
HSS 9.625 × 0.250	25.05	HSS 244.5 × 6.4	37.6
HSS 9.625 × 0.188	18.97	HSS 244.5 × 4.8	28.4
HSS 8.750 × 0.500	44.10	HSS 222.3 × 12.7	65.6
HSS 8.750 × 0.375	33.57	HSS 222.3 × 9.5	49.9
HSS 8.750 × 0.312	28.14	HSS 222.3 × 7.9	41.8
HSS 8.750 × 0.250	22.72	HSS 222.3 × 6.4	34.1
HSS 8.750 × 0.188	17.21	HSS 222.3 × 4.8	25.7
HSS 8.625 × 0.500	43.43	HSS 219.1 × 12.7	64.6
HSS 8.625 × 0.375	33.07	HSS 219.1 × 9.5	49.1
HSS 8.625 × 0.312	28.58	HSS 219.1 × 8.2	42.6

STRUCTURAL STEEL WEIGHT TABLE (IMPERIAL AND METRIC) *(cont.)*

HSS (HOLLOW STRUCTURAL SECTIONS) ROUND

Imperial		Metric	
Designation	**Weight (lb/ft)**	**Designation**	**Weight (kg/m)**
HSS 8.625 × 0.250	22.38	HSS 219.1 × 6.4	33.6
HSS 8.625 × 0.188	16.96	HSS 219.1 × 4.8	25.4
HSS 7.625 × 0.375	29.06	HSS 193.7 × 9.5	43.2
HSS 7.625 × 0.328	25.59	HSS 193.7 × 8.3	37.9
HSS 7.625 × 0.125	10.02	HSS 193.7 × 3.2	15.0
HSS 7.500 × 0.500	37.42	HSS 190.5 × 12.7	55.7
HSS 7.500 × 0.375	28.56	HSS 190.5 × 9.5	42.4
HSS 7.500 × 0.312	23.97	HSS 190.5 × 7.9	35.6
HSS 7.500 × 0.250	19.38	HSS 190.5 × 6.4	29.1
HSS 7.500 × 0.188	14.70	HSS 190.5 × 4.8	22.0
HSS 7.000 × 0.500	34.74	HSS 177.8 × 12.7	51.7
HSS 7.000 × 0.375	26.56	HSS 177.8 × 9.5	39.4
HSS 7.000 × 0.312	22.31	HSS 177.8 × 7.9	33.1
HSS 7.000 × 0.250	18.04	HSS 177.8 × 6.4	27.1
HSS 7.000 × 0.188	13.69	HSS 177.8 × 4.8	20.5
HSS 7.000 × 0.125	9.19	HSS 177.8 × 3.2	13.8
HSS 6.875 × 0.500	34.07	HSS 174.6 × 12.7	50.7
HSS 6.875 × 0.375	26.06	HSS 174.6 × 9.5	38.7
HSS 6.875 × 0.312	21.89	HSS 174.6 × 7.9	32.5
HSS 6.875 × 0.250	17.71	HSS 174.6 × 6.4	26.5

STRUCTURAL STEEL WEIGHT TABLE
(IMPERIAL AND METRIC) *(cont.)*

HSS (HOLLOW STRUCTURAL SECTIONS) ROUND

Imperial		Metric	
Designation	Weight (lb/ft)	Designation	Weight (kg/m)
HSS 6.875 × 0.188	13.44	HSS 174.6 × 4.8	20.1
HSS 6.625 × 0.500	32.74	HSS 168.3 × 12.7	48.7
HSS 6.625 × 0.432	28.60	HSS 168.3 × 11	42.7
HSS 6.625 × 0.375	25.05	HSS 168.3 × 9.5	37.2
HSS 6.625 × 0.312	21.06	HSS 168.3 × 7.9	31.3
HSS 6.625 × 0.280	18.99	HSS 168.3 × 7.1	28.2
HSS 6.625 × 0.250	17.04	HSS 168.3 × 6.4	25.6
HSS 6.625 × 0.188	12.94	HSS 168.3 × 4.8	19.4
HSS 6.625 × 0.125	8.69	HSS 168.3 × 3.2	13.0
HSS 6.125 × 0.500	30.07	HSS 155.6 × 12.7	44.8
HSS 6.125 × 0.375	23.05	HSS 155.6 × 9.5	34.2
HSS 6.125 × 0.312	19.39	HSS 155.6 × 7.9	28.8
HSS 6.125 × 0.250	15.70	HSS 155.6 × 6.4	23.5
HSS 6.125 × 0.188	11.93	HSS 155.6 × 4.8	17.9
HSS 6.000 × 0.500	29.40	HSS 152.4 × 12.7	43.8
HSS 6.000 × 0.375	22.55	HSS 152.4 × 9.5	33.5
HSS 6.000 × 0.312	18.97	HSS 152.4 × 7.9	28.2
HSS 6.000 × 0.280	17.12	HSS 152.4 × 7.1	25.4
HSS 6.000 × 0.250	15.37	HSS 152.4 × 6.4	23.0
HSS 6.000 × 0.188	11.68	HSS 152.4 × 4.8	17.5

STRUCTURAL STEEL WEIGHT TABLE (IMPERIAL AND METRIC) *(cont.)*

HSS (HOLLOW STRUCTURAL SECTIONS) ROUND

Imperial		Metric	
Designation	Weight (lb/ft)	Designation	Weight (kg/m)
HSS 6.000 × 0.125	7.85	HSS 152.4 × 3.2	11.8
HSS 5.563 × 0.375	20.80	HSS 141.3 × 9.5	30.9
HSS 5.563 × 0.258	14.63	HSS 141.3 × 6.6	21.9
HSS 5.563 × 0.188	10.80	HSS 141.3 × 4.8	16.2
HSS 5.563 × 0.134	7.78	HSS 141.3 × 3.4	11.6
HSS 5.500 × 0.500	26.73	HSS 139.7 × 12.7	39.8
HSS 5.500 × 0.375	20.54	HSS 139.7 × 9.5	30.5
HSS 5.500 × 0.258	14.46	HSS 139.7 × 6.6	21.7
HSS 5.000 × 0.500	24.05	HSS 127.0 × 12.7	35.8
HSS 5.000 × 0.375	18.54	HSS 127.0 × 9.5	27.5
HSS 5.000 × 0.312	15.64	HSS 127.0 × 7.9	23.2
HSS 5.000 × 0.280	13.08	HSS 127.0 × 6.6	19.6
HSS 5.000 × 0.250	12.69	HSS 127.0 × 6.4	19.0
HSS 5.000 × 0.188	9.67	HSS 127.0 × 4.8	14.5
HSS 5.000 × 0.125	6.51	HSS 127.0 × 3.2	9.8
HSS 4.500 × 0.337	15.00	HSS 114.3 × 8.6	22.4
HSS 4.500 × 0.237	10.80	HSS 114.3 × 6.0	16.0
HSS 4.500 × 0.188	8.67	HSS 114.3 × 4.8	13.0
HSS 4.500 × 0.125	5.85	HSS 114.3 × 3.2	8.8
HSS 4.000 × 0.337	13.20	HSS 101.6 × 8.6	19.7

STRUCTURAL STEEL WEIGHT TABLE
(IMPERIAL AND METRIC) *(cont.)*

HSS (HOLLOW STRUCTURAL SECTIONS) ROUND

Imperial		Metric	
Designation	Weight (lb/ft)	Designation	Weight (kg/m)
HSS 4.000 × 0.313	12.34	HSS 114.3 × 8.0	18.5
HSS 4.000 × 0.250	10.02	HSS 114.3 × 6.4	15.0
HSS 4.000 × 0.237	9.53	HSS 114.3 × 6.0	14.1
HSS 4.000 × 0.226	9.12	HSS 114.3 × 5.7	13.5
HSS 4.000 × 0.220	8.89	HSS 114.3 × 5.6	13.3
HSS 4.000 × 0.188	7.66	HSS 114.3 × 4.8	11.5
HSS 4.000 × 0.125	5.18	HSS 114.3 × 3.2	7.8
HSS 3.500 × 0.313	10.66	HSS 88.9 × 8.0	16.0
HSS 3.500 × 0.300	10.26	HSS 88.9 × 7.6	15.2
HSS 3.500 × 0.250	8.69	HSS 88.9 × 6.4	13.0
HSS 3.500 × 0.216	7.58	HSS 88.9 × 5.5	11.3
HSS 3.500 × 0.203	7.15	HSS 88.9 × 5.2	10.7
HSS 3.500 × 0.188	6.66	HSS 88.9 × 4.8	10.0
HSS 3.500 × 0.125	4.51	HSS 88.9 × 3.2	6.8
HSS 3.000 × 0.300	8.66	HSS 76.2 × 7.6	12.9
HSS 3.000 × 0.250	7.35	HSS 76.2 × 6.4	11.0
HSS 3.000 × 0.216	6.43	HSS 76.2 × 5.5	9.6
HSS 3.000 × 0.203	6.07	HSS 76.2 × 5.2	9.1

STRUCTURAL STEEL WEIGHT TABLE (IMPERIAL AND METRIC) *(cont.)*

HSS (HOLLOW STRUCTURAL SECTIONS) ROUND

Imperial		Metric	
Designation	**Weight (lb/ft)**	**Designation**	**Weight (kg/m)**
HSS 3.000 × 0.188	5.65	HSS 76.2 × 4.8	8.5
HSS 3.000 × 0.152	4.63	HSS 76.2 × 3.9	7.0
HSS 3.000 × 0.134	4.11	HSS 76.2 × 3.4	6.1
HSS 3.000 × 0.120	3.69	HSS 76.2 × 3.0	5.4
HSS 2.875 × 0.250	7.02	HSS 73.0 × 6.4	10.5
HSS 2.875 × 0.203	5.80	HSS 73.0 × 5.2	8.7
HSS 2.875 × 0.188	5.40	HSS 73.0 × 4.8	8.1
HSS 2.875 × 0.125	3.67	HSS 73.0 × 3.2	5.5
HSS 2.500 × 0.250	6.01	HSS 63.5 × 6.4	9.0
HSS 2.500 × 0.188	4.65	HSS 63.5 × 4.8	6.9
HSS 2.500 × 0.125	3.17	HSS 63.5 × 3.2	4.8
HSS 2.375 × 0.250	5.68	HSS 60.3 × 6.4	8.5
HSS 2.375 × 0.218	5.03	HSS 60.3 × 5.5	7.4
HSS 2.375 × 0.188	4.40	HSS 60.3 × 4.8	6.6
HSS 2.375 × 0.154	3.66	HSS 60.3 × 3.9	5.4
HSS 2.375 × 0.125	3.01	HSS 60.3 × 3.2	4.5
HSS 1.900 × 0.145	2.72	HSS 48.3 × 3.7	4.1
HSS 1.660 × 0.140	2.27	HSS 42.2 × 3.6	3.4

STRUCTURAL STEEL WEIGHT TABLE
(IMPERIAL AND METRIC) *(cont.)*

PIPE (STEEL PIPE)

Imperial		Metric	
Designation	**Weight (lb/ft)**	**Designation**	**Weight (kg/m)**
Standard			
PIPE ½ STD	0.852	PIPE 13 STD	1.24
PIPE ¾ STD	1.13	PIPE 19 STD	1.65
PIPE 1 STD	1.68	PIPE 25 STD	2.45
PIPE 1¼ STD	2.27	PIPE 32 STD	3.32
PIPE 1½ STD	2.72	PIPE 38 STD	3.97
PIPE 2 STD	3.66	PIPE 51 STD	5.34
PIPE 2½ STD	5.8	PIPE 64 STD	8.46
PIPE 3 STD	7.58	PIPE 75 STD	11.10
PIPE 3½ STD	9.12	PIPE 89 STD	13.30
PIPE 4 STD	10.8	PIPE 102 STD	15.80
PIPE 5 STD	14.6	PIPE 127 STD	21.40
PIPE 6 STD	19	PIPE 152 STD	27.70
PIPE 8 STD	28.6	PIPE 203 STD	41.70
PIPE 10 STD	40.5	PIPE 254 STD	59.10
PIPE 12 STD	49.6	PIPE 310 STD	72.40
Extra Strong			
PIPE ½ XS	1.09	PIPE 13 XS	1.59
PIPE ¾ XS	1.48	PIPE 19 XS	2.15
PIPE 1 XS	2.17	PIPE 25 XS	3.17
PIPE 1¼ XS	3	PIPE 32 XS	4.38

STRUCTURAL STEEL WEIGHT TABLE (IMPERIAL AND METRIC) *(cont.)*

PIPE (STEEL PIPE)

Imperial		Metric	
Designation	**Weight (lb/ft)**	**Designation**	**Weight (kg/m)**
Extra Strong (cont.)			
PIPE 1½ XS	3.63	PIPE 38 XS	5.30
PIPE 2 XS	5.03	PIPE 51 XS	7.34
PIPE 2½ XS	7.67	PIPE 64 XS	11.20
PIPE 3 XS	10.3	PIPE 75 XS	15.00
PIPE 3½ XS	12.5	PIPE 89 XS	18.30
PIPE 4 XS	15	PIPE 102 XS	21.90
PIPE 5 XS	20.8	PIPE 127 XS	30.40
PIPE 6 XS	28.6	PIPE 152 XS	41.70
PIPE 8 XS	43.4	PIPE 203 XS	63.40
PIPE 10 XS	54.8	PIPE 254 XS	80.00
PIPE 12 XS	65.5	PIPE 310 XS	95.60
Double Extra Strong			
PIPE 2 XXS	9.04	PIPE 51 XXS	13.20
PIPE 2½ XXS	13.7	PIPE 64 XXS	20.00
PIPE 3 XXS	18.6	PIPE 75 XXS	27.10
PIPE 4 XXS	27.6	PIPE 102 XXS	40.20
PIPE 5 XXS	38.6	PIPE 127 XXS	56.30
PIPE 6 XXS	53.2	PIPE 152 XXS	77.70
PIPE 8 XXS	72.5	PIPE 203 XXS	106.00

STRUCTURAL STEEL WEIGHT TABLE (IMPERIAL AND METRIC) *(cont.)*

PL (STEEL PLATE)

Imperial		Metric	
Thickness (in)	**Weight (lb/sf)**	**Thickness (mm)**	**Weight (kg/m²)**
3/16	7.65	4.69	37.48
1/4	10.20	6.25	49.78
5/16	12.75	7.81	62.23
3/8	15.30	9.38	74.68
7/16	17.85	10.94	87.12
1/2	20.40	12.50	99.57
9/16	22.97	14.29	112.09
5/8	25.50	15.63	124.46
11/16	28.05	17.47	136.89
3/4	30.60	18.75	149.35
7/8	35.70	21.88	174.24
1	40.80	25.00	199.13
1 1/4	51.00	31.25	248.92
1 1/2	61.20	37.50	298.70
1 3/4	71.40	43.75	348.48
2	81.60	50.00	398.27

ESTIMATING LUMBER AND PLYWOOD

DIMENSIONAL LUMBER

Nominal Size	Actual Size	Board feet (BF) per linear foot (LF) of lumber
1 × 1	3/4 × ¾	0.083
1 × 2	3/4 × 1-1/2	0.167
1 × 3	3/4 × 2-1/2	0.250
1 × 4	3/4 × 3-1/2	0.333
1 × 6	3/4 × 5-1/2	0.500
1 × 8	3/4 × 7-1/4	0.667
1 × 10	3/4 × 9-1/4	0.833
1 × 12	3/4 × 11-1/4	1.000
2 × 2	1-1/2 × 1-1/2	0.333
2 × 3	1-1/2 × 2-1/2	0.500
2 × 4	1-1/2 × 3-1/2	0.667
2 × 6	1-1/2 × 5-1/2	1.000
2 × 8	1-1/2 × 7-1/4	1.333
2 × 10	1-1/2 × 9-1/4	1.667
2 × 12	1-1/2 × 11-1/4	2.000
2 × 14	1-1/2 × 13-1/4	2.333
3 × 4	2-1/2 × 3-1/2	1.000
3 × 6	2-1/2 × 5-1/2	1.500
3 × 8	2-1/2 × 7-1/4	2.000
3 × 10	2-1/2 × 9-1/4	2.500
3 × 12	2-1/2 × 11-1/4	3.000
3 × 14	2-1/2 × 13-1/4	3.500

ESTIMATING LUMBER AND PLYWOOD *(cont.)*

DIMENSIONAL LUMBER

Nominal Size	Actual Size	Board feet (BF) per linear foot (LF) of lumber
4 × 4	3-1/2 × 3-1/2	1.333
4 × 6	3-1/2 × 5-1/2	2.000
4 × 8	3-1/2 × 7-1/4	2.667
4 × 10	3-1/2 × 9-1/4	3.333
4 × 12	3-1/2 × 11-1/4	4.000
4 × 14	3-1/2 × 13-1/4	4.667
6 × 6	5-1/2 × 5-1/2	3.000
6 × 8	5-1/2 × 7-1/2	4.000
6 × 10	5-1/2 × 9-1/2	5.000
6 × 12	5-1/2 × 11-1/2	6.000
6 × 14	5-1/2 × 13-1/2	7.000
6 × 16	5-1/2 × 15-1/2	8.000
8 × 8	7-1/2 × 7-1/2	5.333
8 × 10	7-1/2 × 9-1/2	6.667
8 × 12	7-1/2 × 11-1/2	8.000
8 × 14	7-1/2 × 13-1/2	9.333
8 × 16	7-1/2 × 15-1/2	10.667
10 × 10	9-1/2 × 9-1/2	8.333
10 × 12	9-1/2 × 11-1/2	10.000
10 × 14	9-1/2 × 13-1/2	11.667
10 × 16	9-1/2 × 15-1/2	13.333
12 × 12	11-1/2 × 11-1/2	12.000
12 × 14	11-1/2 × 13-1/2	14.000
12 × 16	11-1/2 × 15-1/2	16.000

ESTIMATING LUMBER AND PLYWOOD (cont.)

ESTIMATING BOARD FEET IN GIVEN LENGTHS OF LUMBER

Lumber Size	Length (LF)												
	8	10	12	14	16	18	20	22	24	26	28	30
1 × 1	1	1	1	1	1	1	2	2	2	2	2	2
1 × 2	1	2	2	2	3	3	3	4	4	4	5	5
1 × 3	2	3	3	4	4	5	5	6	6	7	7	8
1 × 4	3	3	4	5	5	6	7	7	8	9	9	10
1 × 6	4	5	6	7	8	9	10	11	12	13	14	15
1 × 8	5	7	8	9	11	12	13	15	16	17	19	20
1 × 10	7	8	10	12	13	15	17	18	20	22	23	25
1 × 12	8	10	12	14	16	18	20	22	24	26	28	30
2 × 2	3	3	4	5	5	6	7	7	8	9	9	10
2 × 3	4	5	6	7	8	9	10	11	12	13	14	15
2 × 4	5	7	8	9	11	12	13	15	16	17	19	20
2 × 6	8	10	12	14	16	18	20	22	24	26	28	30
2 × 8	11	13	16	19	21	24	27	29	32	35	37	40
2 × 10	13	17	20	23	27	30	33	37	40	43	47	50

ESTIMATING LUMBER AND PLYWOOD (cont.)

ESTIMATING BOARD FEET IN GIVEN LENGTHS OF LUMBER

Lumber Size	Length (LF)											
	8	10	12	14	16	18	20	22	24	26	28	30
2 × 12	16	20	24	28	32	36	40	44	48	52	56	60
2 × 14	19	23	28	33	37	42	47	51	56	61	65	70
3 × 4	8	10	12	14	16	18	20	22	24	26	28	30
3 × 6	12	15	18	21	24	27	30	33	36	39	42	45
3 × 8	16	20	24	28	32	36	40	44	48	52	56	60
3 × 10	20	25	30	35	40	45	50	55	60	65	70	75
3 × 12	24	30	36	42	48	54	60	66	72	78	84	90
3 × 14	28	35	42	49	56	63	70	77	84	91	98	105
4 × 4	11	13	16	19	21	24	27	29	32	35	37	40
4 × 6	16	20	24	28	32	36	40	44	48	52	56	60
4 × 8	21	27	32	37	43	48	53	59	64	69	75	80
4 × 10	27	33	40	47	53	60	67	73	80	87	93	100
4 × 12	32	40	48	56	64	72	80	88	96	104	112	120
4 × 14	37	47	56	65	75	84	93	103	112	121	131	140

	90	84	78	72	66	60	54	48	42	36	30	24
6 × 6	90	84	78	72	66	60	54	48	42	36	30	24
6 × 8	120	112	104	96	88	80	72	64	56	48	40	32
6 × 10	150	140	130	120	110	100	90	80	70	60	50	40
6 × 12	180	168	156	144	132	120	108	96	84	72	60	48
6 × 14	210	196	182	168	154	140	126	112	98	84	70	56
6 × 16	240	224	208	192	176	160	144	128	112	96	80	64
8 × 8	160	149	139	128	117	107	96	85	75	64	53	43
8 × 10	200	187	173	160	147	133	120	107	93	80	67	53
8 × 12	240	224	208	192	176	160	144	128	112	96	80	64
8 × 14	280	261	243	224	205	187	168	149	131	112	93	75
8 × 16	320	299	277	256	235	213	192	171	149	128	107	85
10 × 10	250	233	217	200	183	167	150	133	117	100	83	67
10 × 12	300	280	260	240	220	200	180	160	140	120	100	80
10 × 14	350	327	303	280	257	233	210	187	163	140	117	93
10 × 16	400	373	347	320	293	267	240	213	187	160	133	107
12 × 12	360	336	312	288	264	240	216	192	168	144	120	96
12 × 14	420	392	364	336	308	280	252	224	196	168	140	112
12 × 16	480	448	416	384	352	320	288	256	224	192	160	128

Note: Quantities are net and approximate. Please add waste.

ESTIMATING LUMBER AND PLYWOOD *(cont.)*

ESTIMATING PLYWOOD SHEATHING

Wall Area (SF)	Sheet Size (LF × LF)		
	4 × 8	4 × 9	4 × 10
100	4	3	3
200	7	6	5
300	10	9	8
400	13	12	10
500	16	14	13
600	19	17	15
700	22	20	18
800	25	23	20
900	29	25	23
1000	32	28	25
1100	35	31	28
1200	38	34	30
1300	41	37	33
1400	44	39	35
1500	47	42	38
1600	50	45	40
1700	54	48	43
1800	57	50	45
1900	60	53	48
2000	63	56	50

Note:
1. Sheet of plywood = Coverage Area/ Sheet Area
2. The most common sheet size is 4' × 8'.
3. Quantities are net and approximate.

ESTIMATING FLOOR JOISTS

ESTIMATING MATH

Joist Spacing	Multiply Floor Length (LF) By	Add
12"	1.00	1
16"	0.75	1
20"	0.60	1
24"	0.50	1
30"	0.40	1
36"	0.33	1
42"	0.29	1
48"	0.25	1
54"	0.22	1
60"	0.20	1

Example:

For 30' long wood floor with joists spacing 24" O.C.
The number of joists is:

$30 \times 0.5 + 1 = 16$ EA

Note: Allow additional numbers of joists where partitions run parallel to joist.

ESTIMATING FLOOR JOISTS (cont.)

NUMBER OF WOOD JOISTS FOR GIVEN FLOOR LENGTH AND JOIST SPACING

Floor Length (LF)	Joist Spacing (in.)										
	12"	16"	20"	24"	30"	36"	42"	48"	54"	60"
6	7	6	5	4	3	3	3	3	2	2
7	8	6	5	5	4	3	3	3	3	2
8	9	7	6	5	4	4	3	3	3	3
9	10	8	6	6	5	4	4	3	3	3
10	11	9	7	6	5	4	4	4	3	3
11	12	9	8	7	5	5	4	4	3	3
12	13	10	8	7	6	5	4	4	4	3
13	14	11	9	8	6	5	5	4	4	4
14	15	12	9	8	7	6	5	5	4	4
15	16	12	10	9	7	6	5	5	4	4
16	17	13	11	9	7	6	6	5	5	4
17	18	14	11	10	8	7	6	5	5	4
18	19	15	12	10	8	7	6	6	5	5
19	20	15	12	11	9	7	7	6	5	5

20	5	5	6	7	8	9	11	13	16	21
21	5	6	6	7	8	9	12	14	17	22
22	5	6	7	7	8	10	12	14	18	23
23	6	6	7	8	9	10	13	15	18	24
24	6	6	7	8	9	11	13	15	19	25
25	6	7	7	8	9	11	14	16	20	26
26	6	7	8	9	10	11	14	17	21	27
27	6	7	8	9	10	12	15	17	21	28
28	7	7	8	9	10	12	15	18	22	29
29	7	7	8	9	11	13	16	18	23	30
30	7	8	9	10	11	13	16	19	24	31
31	7	8	9	10	11	13	17	20	24	32
32	7	8	9	10	12	14	17	20	25	33
33	8	8	9	11	12	14	18	21	26	34
34	8	8	10	11	12	15	18	21	27	35
35	8	9	10	11	13	15	19	22	27	36
36	8	9	10	11	13	15	19	23	28	37
37	8	9	10	12	13	16	20	23	29	38

ESTIMATING FLOOR JOISTS (cont.)

NUMBER OF WOOD JOISTS FOR GIVEN FLOOR LENGTH AND JOIST SPACING (cont.)

Floor Length (LF)	Joist Spacing (in.)										
	12"	16"	20"	24"	30"	36"	42"	48"	54"	60"	
38	39	30	24	20	16	14	12	11	9	9	
39	40	30	24	21	17	14	12	11	10	9	
40	41	31	25	21	17	14	13	11	10	9	
41	42	32	26	22	17	15	13	11	10	9	
42	43	33	26	22	18	15	13	12	10	9	
43	44	33	27	23	18	15	13	12	10	10	
44	45	34	27	23	19	16	14	12	11	10	
45	46	35	28	24	19	16	14	12	11	10	
46	47	36	29	24	19	16	14	13	11	10	
47	48	36	29	25	20	17	15	13	11	10	
48	49	37	30	25	20	17	15	13	12	11	
49	50	38	30	26	21	17	15	13	12	11	
50	51	39	31	26	21	18	16	14	12	11	

Note: Quantities are net and approximate. Please add waste.

ESTIMATING WALL STUDS		
ESTIMATING MATH		
Stud Spacing	**Multiply Partition Length (ft.) By**	**Add**
12"	1.00	1
16"	0.75	1
20"	0.60	1
24"	0.50	1

Example:

For 20' long wood framed wall with studs spacing 16" O.C.
The number of studs is:
$20 \times 0.5 + 1 = 11$ EA

Note: Allow additional number of studs for corners and openings

ESTIMATING WALL STUDS *(cont..)*

NUMBER OF WOOD STUDS FOR GIVEN PARTITION LENGTH AND STUD SPACING

Partition Length (ft.)	Stud Spacing (in.)			
	12"	16"	20"	24"
2	3	3	2	2
3	4	3	3	3
4	5	4	3	3
5	6	5	4	4
6	7	6	5	4
7	8	6	5	5
8	9	7	6	5
9	10	8	6	6
10	11	9	7	6
11	12	9	8	7
12	13	10	8	7
13	14	11	9	8
14	15	12	9	8
15	16	12	10	9
16	17	13	11	9
17	18	14	11	10
18	19	15	12	10
19	20	15	12	11
20	21	16	13	11
21	22	17	14	12
22	23	18	14	12
23	24	18	15	13
24	25	19	15	13
25	26	20	16	14

ESTIMATING WALL STUDS *(cont.)*

NUMBER OF WOOD STUDS FOR GIVEN PARTITION LENGTH AND STUD SPACING *(cont.)*

Partition Length (ft.)	Stud Spacing (in.)			
	12"	16"	20"	24"
26	27	21	17	14
27	28	21	17	15
28	29	22	18	15
29	30	23	18	16
30	31	24	19	16
31	32	24	20	17
32	33	25	20	17
33	34	26	21	18
34	35	27	21	18
35	36	27	22	19
36	37	28	23	19
37	38	29	23	20
38	39	30	24	20
39	40	30	24	21
40	41	31	25	21
41	42	32	26	22
42	43	33	26	22
43	44	33	27	23
44	45	34	27	23
45	46	35	28	24
46	47	36	29	24
47	48	36	29	25
48	49	37	30	25
49	50	38	30	26

NUMBER OF WOOD STUDS FOR GIVEN PARTITION LENGTH AND STUD SPACING *(cont.)*

Partition Length (ft.)	Stud Spacing (in.)			
	12"	16"	20"	24"
50	51	39	31	26
51	52	39	32	27
52	53	40	32	27
53	54	41	33	28
54	55	42	33	28
55	56	42	34	29
56	57	43	35	29
57	58	44	35	30
58	59	45	36	30
59	60	45	36	31
60	61	46	37	31
61	62	47	38	32
62	63	48	38	32
63	64	48	39	33
64	65	49	39	33
65	66	50	40	34
66	67	51	41	34
67	68	51	41	35
68	69	52	42	35
69	70	53	42	36
70	71	54	43	36
71	72	54	44	37
72	73	55	44	37
73	74	56	45	38

ESTIMATING WALL STUDS (cont.)

NUMBER OF WOOD STUDS FOR GIVEN PARTITION LENGTH AND STUD SPACING (cont.)

Partition Length (ft.)	Stud Spacing (in.)			
	12"	16"	20"	24"
74	75	57	45	38
75	76	57	46	39
76	77	58	47	39
77	78	59	47	40
78	79	60	48	40
79	80	60	48	41
80	81	61	49	41
81	82	62	50	42
82	83	63	50	42
83	84	63	51	43
84	85	64	51	43
85	86	65	52	44
86	87	66	53	44
87	88	66	53	45
88	89	67	54	45
89	90	68	54	46
90	91	69	55	46
91	92	69	56	47
92	93	70	56	47
93	94	71	57	48
94	95	72	57	48
95	96	72	58	49
96	97	73	59	49
97	98	74	59	50
98	99	75	60	50
99	100	75	60	51
100	101	76	61	51

ESTIMATING ROOF RAFTER

Estimating Math

Pitch = Rise/Span

Slope = Rise/Run

Total Run = Run Distance + Eavehang Distance

Common Rafter Length= Total Run × Multiplication Factor 1
(See Table Below)

Hip/Valley Rafter Length= Total Run × Multiplication Factor 2
(See Table Below)

Multiplication Factors for Sloped Roof Rafter

Roof Pitch	Roof Slope	Multiply Length of Run By	
		For Common Rafter	For Hip/Valley Rafter
1/12	2 in 12	1.014	1.424
1/8	3 in 12	1.031	1.436
1/6	4 in 12	1.054	1.453
5/24	5 in 12	1.083	1.474
1/4	6 in 12	1.118	1.500
7/24	7 in 12	1.158	1.530
1/3	8 in 12	1.202	1.564
3/8	9 in 12	1.250	1.601
5/12	10 in 12	1.302	1.642
11/24	11 in 12	1.357	1.685
1/2	12 in 12	1.413	1.732
13/24	13 in 12	1.474	1.782
7/12	14 in 12	1.537	1.833
5/8	15 in 12	1.601	1.887
2/3	16 in 12	1.667	1.944
17/24	17 in 12	1.734	2.002
3/4	18 in 12	1.803	2.032
19/24	19 in 12	1.875	2.123
5/6	20 in 12	1.948	2.186
7/8	21 in 12	2.010	2.250
11/12	22 in 12	2.083	2.315
23/24	23 in 12	2.167	2.382
Full	24 in 12	2.240	2.450

ESTIMATING WOOD FLOOR

LAMINATED FLOOR

Lumber Size	Floor Thickness	Linear Foot of Lumber Required Per Square Foot of Floor
2 × 4	1-5/8	4.90
2 × 6	1-5/8	7.40
2 × 8	1-5/8	9.84
2 × 10	1-5/8	12.30
2 × 12	1-5/8	14.76
2 × 14	1-5/8	17.22
3 × 6	2-3/4	6.60
3 × 8	2-3/4	8.80
3 × 10	2-3/4	11.10
3 × 12	2-3/4	13.20
3 × 14	2-3/4	15.40
4 × 6	3-3/4	6.42
4 × 8	3-3/4	8.56
4 × 10	3-3/4	10.70
4 × 12	3-3/4	12.84

HARDWOOD FLOOR

Nominal Size	Finished Size	Add for Waste	Board Foot of Lumber Per 100 Square Feet of Floor
1 × 1	$3/8 × 7/8$	16.67%	117
1 × 2	$3/8 × 1 1/2$	33.33%	133
1 × 2 1/2	$3/8 × 2$	25%	125
1 × 2 1/4	$25/32 × 1 1/2$	50.33%	150
1 × 2 3/4	$25/32 × 2$	37.50%	138
1 × 3	$25/32 × 2 1/4$	33.33%	133
1 × 4	$25/32 × 3 1/4$	25%	125

ESTIMATING ROOFING

ESTIMATING MATH

Flat Roof Area = Roof Length × Roof Width

Sloped Roof Area = Flat Roof Area × Multiplication Factor

Roof Pitch	Roof Slope	Multiply Flat Roof Area By
$1/12$	2 in 12	1.014
$1/8$	3 in 12	1.031
$1/6$	4 in 12	1.054
$5/24$	5 in 12	1.083
$1/4$	6 in 12	1.118
$7/24$	7 in 12	1.158
$1/3$	8 in 12	1.202
$3/8$	9 in 12	1.250
$5/12$	10 in 12	1.302
$11/24$	11 in 12	1.357
$1/2$	12 in 12	1.413
$13/24$	13 in 12	1.474
$7/12$	14 in 12	1.537
$5/8$	15 in 12	1.601
$2/3$	16 in 12	1.667
$17/24$	17 in 12	1.734
$3/4$	18 in 12	1.803
$19/24$	19 in 12	1.875
$5/6$	20 in 12	1.948
$7/8$	21 in 12	2.010
$11/12$	22 in 12	2.083
$23/24$	23 in 12	2.167
Full	24 in 12	2.240

ESTIMATING STUCCO

MATERIALS TO COVER 1 SQ (100 SF) OF STUCCO

Stucco Thickness (in.)	Mortar (CF)	Mix Ratio							
		1:2		1:2.5		1:3		1:3.5	
		Masonry (Bag)	Sand (CF)	Masonry (Bag)	Sand (CF)	Masonry (Bag)	Sand (CF)	Masonry (Bag)	Sand (CF)
1/4"	2.1	0.95	1.9	0.79	2	0.69	2.1	0.59	2.1
3/8"	3.1	1.41	2.8	1.17	2.9	1.01	3	0.87	3
1/2"	4.2	1.91	3.8	1.58	4	1.37	4.1	1.18	4.1
5/8"	5.2	2.36	4.7	1.96	4.9	1.7	5.1	1.47	5.2
3/4"	6.3	2.86	5.7	2.38	5.9	2.06	6.2	1.78	6.2
1"	8.3	3.77	7.5	3.13	7.8	2.71	8.1	2.34	8.2

ESTIMATING SIDING

Estimating Math

1. Siding for Rectangular Area = Wall Length × Wall Height (to bottom of gable)
2. Siding for Gable Area = Wall Length × Gable Height × 1/2
3. Total Siding = Siding for Rectangular Area + Siding for Gable Area

Gable Area Table

Wall Length	Gable Height							
	3'	4'	5'	6'	7'	8'	9'	10'
4'	6	8	10	12	14	16	18	20
8'	12	16	20	24	28	32	36	40
12'	18	24	30	36	42	48	54	60
16'	24	32	40	48	56	64	72	80
20'	30	40	50	60	70	80	90	100
24'	36	48	60	72	84	96	108	120
28'	42	56	70	84	98	112	126	140
32'	48	64	80	96	112	128	144	160
36'	54	72	90	108	126	144	162	180
40'	60	80	100	120	140	160	180	200
44'	66	88	110	132	154	176	198	220
48'	72	96	120	144	168	192	216	240
52'	78	104	130	156	182	208	234	260
56'	84	112	140	168	196	224	252	280
60'	90	120	150	180	210	240	270	300

Example:

The building elevation shows wall with siding is 44' long and 15' high. The gable is an additional 6' high.

Rectangular Area = 44' × 15' = 660 SF

Gable Area = 132 SF (from the table)

Total Siding Area = 660 + 132 = 792 SF

Add 10% waste: 792 × (1 + 10%) = 872 SF

ESTIMATING DRYWALL

Estimating Math

Wall Area (SF) = Wall Length $\times$ Wall Width

Quantity of Wall Boards (EA) = Wall Area $\times$ Layers of Drywall / Board Area

Ready Mix (LB) = Wall Area $\times$ 0.14

Tape (LF) = Wall Area $\times$ 0.48

Joint Compounds (LB) = Wall Area $\times$ 0.09

Nails (LB) = Wall Area $\times$ 0.005

Screws (EA) = Wall Area $\times$ 1.25

Note:
1. Drywall for ceiling area is similar.
2. EA = Each, LB = Pounds, LF = Linear Foot,
   SF = Square Foot

For Interior Room Finish (Rectangular):

Wall Finish = (Room Length + Room Width) $\times$
2 $\times$ Room Height

Ceiling Finish = Room Length $\times$ Room Width

ESTIMATING DRYWALL *(cont.)*

Quantities of Drywall and Accessories for Given Wall/Ceiling Areas

Area (SF)	Wallboard Quantities			Ready Mix (LB)	Joint Tape (LF)	Joint Compound (LB)	Nail (LB)	Screw (EA)
	4'×8'	4'×10'	4'×12'					
100	4	3	3	14	48	9	0.5	125
200	7	5	5	28	96	18	1.0	250
300	10	8	7	42	144	27	1.5	375
400	13	10	9	56	192	36	2.0	500
500	16	13	11	70	240	45	2.5	625
600	19	15	13	84	288	54	3.0	750
700	22	18	15	98	336	63	3.5	875
800	25	20	17	112	384	72	4.0	1000
900	29	23	19	126	432	81	4.5	1125
1000	32	25	21	140	480	90	5.0	1250
1100	35	28	23	154	528	99	5.5	1375
1200	38	30	25	168	576	108	6.0	1500
1300	41	33	28	182	624	117	6.5	1625
1400	44	35	30	196	672	126	7.0	1750
1500	47	38	32	210	720	135	7.5	1875
1600	50	40	34	224	768	144	8.0	2000
1700	54	43	36	238	816	153	8.5	2125
1800	57	45	38	252	864	162	9.0	2250
1900	60	48	40	266	912	171	9.5	2375
2000	63	50	42	280	960	180	10.0	2500
2500	79	63	53	350	1200	225	12.5	3125
3000	94	75	63	420	1440	270	15.0	3750
3500	110	88	73	490	1680	315	17.5	4375
4000	125	100	84	560	1920	360	20.0	5000
4500	141	113	94	630	2160	405	22.5	5625
5000	157	125	105	700	2400	450	25.0	6250

Note:
1. The above table only includes single layer of drywall boards.
2. Quantities are net and approximate. Please add waste.

ESTIMATING DRYWALL *(cont.)*				
NUMBER OF WALL BOARDS FOR ROOMS 8' HIGH				
Room Size	**4 × 8 Board**	**4 × 9 Board**	**4 × 10 Board**	**4 × 12 Board**
8' × 8'	8	8	7	6
8' × 9'	9	8	7	6
8' × 10'	9	8	8	6
8' × 11'	10	9	8	7
8' × 12'	10	9	8	7
8' × 13'	11	10	9	7
8' × 14'	11	10	9	8
8' × 15'	12	11	10	8
8' × 16'	12	11	10	8
9' × 9'	9	8	8	6
9' × 10'	10	9	8	7
9' × 11'	10	9	8	7
9' × 12'	11	10	9	7
9' × 13'	11	10	9	8
9' × 14'	12	11	10	8
9' × 15'	12	11	10	8
9' × 16'	13	12	10	9
10' × 10'	10	9	8	7
10' × 11'	11	10	9	7
10' × 12'	11	10	9	8
10' × 13'	12	11	10	8
10' × 14'	12	11	10	8
10' × 15'	13	12	10	9
10' × 16'	13	12	11	9

ESTIMATING DRYWALL *(cont.)*

NUMBER OF WALL BOARDS FOR ROOMS 8' HIGH

Room Size	4 × 8 Board	4 × 9 Board	4 × 10 Board	4 × 12 Board
11' × 11'	11	10	9	8
11' × 12'	12	11	10	8
11' × 13'	12	11	10	8
11' × 14'	13	12	10	9
11' × 15'	13	12	11	9
11' × 16'	14	12	11	9
12' × 12'	12	11	10	8
12' × 13'	13	12	10	9
12' × 14'	13	12	11	9
12' × 15'	14	12	11	9
12' × 16'	14	13	12	10
13' × 13'	13	12	11	9
13' × 14'	14	12	11	9
13' × 15'	14	13	12	10
13' × 16'	15	13	12	10
14' × 14'	14	13	12	10
14' × 15'	15	13	12	10
14' × 16'	15	14	12	10
15' × 15'	15	14	12	10
15' × 16'	16	14	13	11
16' × 16'	16	15	13	11

Note:
1. Single layer of wall boards on interior side of room only.
   No ceiling boards included.
2. Quantities are net and approximate. Please add waste.

ESTIMATING DRYWALL *(cont.)*				
NUMBER OF WALL BOARDS FOR ROOMS 9' HIGH				
Room Size	4 × 8 Board	4 × 9 Board	4 × 10 Board	4 × 12 Board
8' × 8'	9	8	8	6
8' × 9'	10	9	8	7
8' × 10'	11	9	9	7
8' × 11'	11	10	9	8
8' × 12'	12	10	9	8
8' × 13'	12	11	10	8
8' × 14'	13	11	10	9
8' × 15'	13	12	11	9
8' × 16'	14	12	11	9
9' × 9'	11	9	9	7
9' × 10'	11	10	9	8
9' × 11'	12	10	9	8
9' × 12'	12	11	10	8
9' × 13'	13	11	10	9
9' × 14'	13	12	11	9
9' × 15'	14	12	11	9
9' × 16'	15	13	12	10
10' × 10'	12	10	9	8
10' × 11'	12	11	10	8
10' × 12'	13	11	10	9
10' × 13'	13	12	11	9
10' × 14'	14	12	11	9
10' × 15'	15	13	12	10
10' × 16'	15	13	12	10

ESTIMATING DRYWALL *(cont.)*				
NUMBER OF WALL BOARDS FOR ROOMS 9' HIGH				
Room Size	**4 × 8 Board**	**4 × 9 Board**	**4 × 10 Board**	**4 × 12 Board**
11' × 11'	13	11	10	9
11' × 12'	13	12	11	9
11' × 13'	14	12	11	9
11' × 14'	15	13	12	10
11' × 15'	15	13	12	10
11' × 16'	16	14	13	11
12' × 12'	14	12	11	9
12' × 13'	15	13	12	10
12' × 14'	15	13	12	10
12' × 15'	16	14	13	11
12' × 16'	16	14	13	11
13' × 13'	15	13	12	10
13' × 14'	16	14	13	11
13' × 15'	16	14	13	11
13' × 16'	17	15	14	11
14' × 14'	16	14	13	11
14' × 15'	17	15	14	11
14' × 16'	17	15	14	12
15' × 15'	17	15	14	12
15' × 16'	18	16	14	12
16' × 16'	18	16	15	12

Note:
1. Single layer of wall boards on interior side of room only.
   No ceiling boards included.
2. Quantities are net and approximate. Please add waste.

ESTIMATING DRYWALL *(cont.)*				
NUMBER OF WALL BOARDS FOR ROOMS 10' HIGH				
Room Size	4 × 8 Board	4 × 9 Board	4 × 10 Board	4 × 12 Board
8' × 8'	10	9	8	7
8' × 9'	11	10	9	8
8' × 10'	12	10	9	8
8' × 11'	12	11	10	8
8' × 12'	13	12	10	9
8' × 13'	14	12	11	9
8' × 14'	14	13	11	10
8' × 15'	15	13	12	10
8' × 16'	15	14	12	10
9' × 9'	12	10	9	8
9' × 10'	12	11	10	8
9' × 11'	13	12	10	9
9' × 12'	14	12	11	9
9' × 13'	14	13	11	10
9' × 14'	15	13	12	10
9' × 15'	15	14	12	10
9' × 16'	16	14	13	11
10' × 10'	13	12	10	9
10' × 11'	14	12	11	9
10' × 12'	14	13	11	10
10' × 13'	15	13	12	10
10' × 14'	15	14	12	10
10' × 15'	16	14	13	11
10' × 16'	17	15	13	11

ESTIMATING DRYWALL *(cont.)*

NUMBER OF WALL BOARDS FOR ROOMS 10' HIGH

Room Size	4 × 8 Board	4 × 9 Board	4 × 10 Board	4 × 12 Board
11' × 11'	14	13	11	10
11' × 12'	15	13	12	10
11' × 13'	15	14	12	10
11' × 14'	16	14	13	11
11' × 15'	17	15	13	11
11' × 16'	17	15	14	12
12' × 12'	15	14	12	10
12' × 13'	16	14	13	11
12' × 14'	17	15	13	11
12' × 15'	17	15	14	12
12' × 16'	18	16	14	12
13' × 13'	17	15	13	11
13' × 14'	17	15	14	12
13' × 15'	18	16	14	12
13' × 16'	19	17	15	13
14' × 14'	18	16	14	12
14' × 15'	19	17	15	13
14' × 16'	19	17	15	13
15' × 15'	19	17	15	13
15' × 16'	20	18	16	13
16' × 16'	20	18	16	14

Note:
1. Single layer of wall boards on interior side of room only. No ceiling boards included.
2. Quantities are net and approximate. Please add waste.

ESTIMATING DRYWALL *(cont.)*				
NUMBER OF CEILING BOARDS				
Room Size	**4 × 8 Board**	**4 × 9 Board**	**4 × 10 Board**	**4 × 12 Board**
8' × 8'	2	2	2	2
8' × 9'	3	2	2	2
8' × 10'	3	3	2	2
8' × 11'	3	3	3	2
8' × 12'	3	3	3	2
8' × 13'	4	3	3	3
8' × 14'	4	4	3	3
8' × 15'	4	4	3	3
8' × 16'	4	4	4	3
9' × 9'	3	3	3	2
9' × 10'	3	3	3	2
9' × 11'	4	3	3	3
9' × 12'	4	3	3	3
9' × 13'	4	4	3	3
9' × 14'	4	4	4	3
9' × 15'	5	4	4	3
9' × 16'	5	4	4	3
10' × 10'	4	3	3	3
10' × 11'	4	4	3	3
10' × 12'	4	4	3	3
10' × 13'	5	4	4	3
10' × 14'	5	4	4	3
10' × 15'	5	5	4	4
10' × 16'	5	5	4	4

ESTIMATING DRYWALL (cont.)

NUMBER OF CEILING BOARDS

Room Size	4 × 8 Board	4 × 9 Board	4 × 10 Board	4 × 12 Board
11' × 11'	4	4	4	3
11' × 12'	5	4	4	3
11' × 13'	5	4	4	3
11' × 14'	5	5	4	4
11' × 15'	6	5	5	4
11' × 16'	6	5	5	4
12' × 12'	5	4	4	3
12' × 13'	5	5	4	4
12' × 14'	6	5	5	4
12' × 15'	6	5	5	4
12' × 16'	6	6	5	4
13' × 13'	6	5	5	4
13' × 14'	6	6	5	4
13' × 15'	7	6	5	5
13' × 16'	7	6	6	5
14' × 14'	7	6	5	5
14' × 15'	7	6	6	5
14' × 16'	7	7	6	5
15' × 15'	8	7	6	5
15' × 16'	8	7	6	5
16' × 16'	8	8	7	6

Note:
1. Ceiling boards only. No wall boards included.
2. Quantities are net and approximate. Please add waste.

ESTIMATING PAINTING
Estimating Math

For Room Interior Finish:

Wall to be painted = (Room Length + Room Width) × 2 × Room Height

Ceiling to be painted = Room Length × Room Width

Doors/Windows to be painted = Door/Window Height × Door/Window Width

Wood Trims to be painted = Trim Length × Trim Width

Surface to be painted = Walls + Ceilings + Door/Window + Trim

Amount of paint needed for each coat = Surface to be painted/ Coverage Rate

Coverage Rate for Exterior Paint (Square Feet per Gallon)

Surface and Paint Type	First Coat	Second Coat	Third Coat
Brick (oil base)	200	400	N/A
Brick (water base)	100	150	N/A
Cement floor (paint)	450	600	600
Cement floor (stain)	510	480	N/A
Gutter	200	N/A	N/A
Porch floor (wood)	378	540	576
Siding, shingle (paint)	342	423	N/A
Siding, shingle (stain)	150	225	N/A
Siding (paint)	420	520	620
Stucco, medium texture (oil)	153	360	360
Stucco, medium texture (water/cement)	99	135	N/A
Trim	850	900	972

Coverage Rate for Interior Paint (Square Feet per Gallon)

Surface and Paint Type	First Coat
Concrete block	200
Gypsum board (flat)	450
Gypsum board (gloss)	400
Plaster (texture)	250

ESTIMATING FLOORING

Estimating Math

For interior room finish (Rectangular)

Floor Finish = Room Length × Room Width

Floor Base = (Room Length + Room Width) × 2 × Base Height

For irregularly shaped floors, divide them into individual sections, calculate the square feet in each one, and then add them together.

Tile Size	Number of Tiles per Square Foot
3" × 3"	16
4" × 4"	9
4¼" × 4 ¼"	8
6" × 6"	4
6" × 8"	3
6" × 12"	2
8" × 8"	2.25
9" × 9"	1.78
10" × 10"	1.45
12" × 12"	1

Note: Quantities are net and approximate. Please add waste.

PLUMBING PIPES

Pipe Types	Application
Copper Tubing	
Type K (Green)	Heaviest copper tubing. Mainly used for underground such as domestic water services, fire protection, heating, steam, gas, oil, oxygen, etc. Available in both hard and soft types.
Type L (Blue)	Standard tubing used for interior, above ground plumbing or underground drainage. Available in both hard and soft types.
Type M (Red)	Lightest copper tubing. Interior heating and pressure applications, thickness less than type K and L. Used for most residential work. Available in both hard and soft types.
Type DWV (Yellow)	"Drain, Waste, Vent". Above ground use only and no pressure applications. Available only in hard type.
Steel Pipe	
Black Pipe	Painted and has very little resistance to rust
Galvanized Pipe	Coated with zinc inside and out and has very good resistance to rust
Plastic Pipe	
PVC	Cold water or waste water applications but not for hot water systems.
CPVC	Hot water applications
ABS	Underground applications like sanitary drainage, vent piping, storm water drainage
PE	Pressurized water system such as sprinklers. Cold water applications only.
PB	Pressurized water system. Both hot and cold water applications.
Others	
Malleable Iron Pipe	Drain pipes, hot and cold supply
Cast Iron Pipe	Sanitary and storm drain, waste and vent piping applications
Vitrified Clay Pipe	Sanitary and storm drain, waste and vent piping applications

PLUMBING PIPES (cont.)

COPPER TUBING PROPERTIES

Nominal Size (in)	Actual Outside Diameter Sizes (in)	Type K		Type L		Type M		Type DWV	
		Wall Thickness (in.)	Weight (lb/ft)	Wall Thickness (in.)	Weight (lb/ft)	Wall Thickness (in.)	Weight (lb/ft)	Wall Thickness (in.)	Weight (lb/ft)
1/4	0.375	0.035	0.145	0.030	0.126	N/A	N/A	N/A	N/A
3/8	0.5	0.049	0.269	0.035	0.198	0.025	0.145	N/A	N/A
1/2	0.625	0.049	0.344	0.040	0.285	0.028	0.204	N/A	N/A
5/8	0.75	0.049	0.418	0.042	0.362	N/A	N/A	N/A	N/A
3/4	0.875	0.065	0.641	0.045	0.455	0.032	0.328	N/A	N/A
1	1.125	0.065	0.839	0.050	0.655	0.035	0.465	N/A	N/A
1 1/4	1.375	0.065	1.040	0.055	0.884	0.042	0.682	0.040	0.65
1 1/2	1.625	0.072	1.360	0.060	1.140	0.049	0.940	0.042	0.81
2	2.125	0.083	2.060	0.070	1.750	0.058	1.460	0.042	1.07
2 1/2	2.625	0.095	2.930	0.080	2.480	0.065	2.030	N/A	N/A
3	3.125	0.109	4.000	0.090	3.330	0.072	2.680	0.045	1.69
3 1/2	3.625	0.120	5.120	0.100	4.290	0.083	3.580	N/A	N/A
4	4.125	0.134	6.510	0.110	5.380	0.095	4.660	0.058	2.87
5	5.125	0.160	9.670	0.125	7.610	0.109	6.660	0.072	4.43
6	6.125	0.192	13.90	0.140	10.20	0.122	8.92	0.083	6.10
8	8.125	0.271	25.90	0.200	19.30	0.170	16.50	N/A	N/A
10	10.125	0.338	40.30	0.250	30.10	0.212	25.60	N/A	N/A
12	12.125	0.405	57.80	0.280	40.40	0.254	36.70	N/A	N/A

PLUMBING PIPES (cont.)

PLASTIC PIPE PROPERTIES

Nominal Size (in)	Actual OD Sizes (in)	PVC Schedule 40		PVC Schedule 80		CPVC Schedule 40		CPVC Schedule 80	
		Wall Thickness (in.)	Weight (lb/ft)	Wall Thickness (in.)	Weight (lb/ft)	Wall Thickness (in.)	Weight (lb/ft)	Wall Thickness (in.)	Weight (lb/ft)
1/4	0.540	N/A	N/A	0.119	0.10	N/A	N/A	0.119	0.12
1/2	0.840	0.109	0.16	0.147	0.21	0.109	0.19	0.147	0.24
3/4	1.050	0.113	0.22	0.154	0.28	0.113	0.25	0.154	0.33
1	1.315	0.133	0.32	0.179	0.40	0.133	0.38	0.179	0.49
1 1/4	1.660	0.140	0.43	0.191	0.57	0.140	0.51	0.191	0.67
1 1/2	1.900	0.145	0.52	0.200	0.69	0.145	0.61	0.200	0.81
2	2.375	0.154	0.70	0.218	0.95	0.154	0.82	0.218	1.09
2 1/2	2.875	0.203	1.10	0.276	1.45	0.203	1.29	0.276	1.65
3	3.500	0.216	1.44	0.300	1.94	0.216	1.69	0.300	2.21
4	4.500	0.237	2.05	0.337	2.83	0.237	2.33	0.337	3.23
6	6.625	0.280	3.61	0.432	5.41	0.280	4.10	0.432	6.17
8	8.625	0.322	5.45	0.500	8.22	N/A	N/A	0.500	9.06
10	10.750	0.365	7.91	0.593	12.28	N/A	N/A	N/A	N/A
12	12.750	0.406	10.35	0.687	17.10	N/A	N/A	N/A	N/A

PLUMBING PIPES (cont.)				
STANDARD STEEL PIPE PROPERTIES				
Nominal Size (in)	Actual Outside Diameter Sizes (in)	Wall Thickness (in)	Internal Area (sq in)	Weight (lb/ft)
1/8	0.405	0.068	0.057	0.24
1/4	0.540	0.088	0.104	0.42
3/8	0.675	0.091	0.191	0.56
1/2	0.840	0.109	0.304	0.84
3/4	1.050	0.113	0.533	1.12
1	1.315	0.133	0.861	1.67
1 1/4	1.660	0.140	1.496	2.25
1 1/2	1.900	0.145	2.036	2.68
2	2.375	0.154	3.356	3.61
2 1/2	2.875	0.203	4.780	5.74
3	3.500	0.217	7.383	7.54
3 1/2	4.000	0.226	9.886	9.00
4	4.500	0.237	12.730	10.67
5	5.563	0.259	19.985	14.50
6	6.625	0.280	28.886	18.76
7	7.625	0.301	38.734	23.27
8	8.625	0.322	50.021	28.18
9	9.625	0.344	62.720	33.70
10	10.750	0.336	78.820	40.07
12	12.750	0.375	113.090	48.99

PLUMBING PIPES (cont.)

PRESSURE RATING OF SCHEDULE 40 STEEL PIPE

Pipe	Pressure
1/8 to 1 inch continuous weld or seamless	700 psi
1 1/4 to 3 inch continuous weld	800 psi
3 1/2 to 4 inch continuous weld	1200 psi
2 to 12 inch electric weld	1000 to 1300 psi
1 1/4 to 3 inch seamless	1000 psi
3 to 12 inch seamless	1000 to 1300 psi

HVAC DUCTWORK		
Ductwork Gauge	**Weight (lb/sf)**	**Section Long Side Max Dimension (in)**
26	0.906	N/A
24	1.156	30
22	1.406	54
20	1.656	84
18	2.156	85 and up
16	2.656	N/A

Estimating Example:
Find the weight for 34" × 20" ductwork 15' long

Calculation:
Section long side dimension is 34", more than 30" but less than 54"
So this is a 22 gauge ductwork
Area = (34 + 20) × 2/12 × 15 = 135 SF

Add 10% waste:
Area = 135 × (1 + 10%) = 149 SF
Weight = 149 SF × 1.406 LBS/SF = 210 LBS

ELECTRICAL WIRES AND CONDUITS

WIRE SIZES

Size (AWG)	Cir Mills (area)	Sq MM (area)
18	1,620	0.82
16	2,580	1.30
14	4,110	2.08
12	6,530	3.30
10	10,380	5.25
8	16,510	8.36
6	26,240	13.29
4	41,740	21.14
3	52,620	26.65
2	66,360	33.61
1	83,690	42.39
1/0	105,600	53.49
2/0	133,100	67.42
3/0	167,800	85.00
4/0	211,600	107.19
250	250,000	126.64
300	300,000	151.97
350	350,000	177.3
400	400,000	202.63
500	500,000	253.29

ELECTRICAL WIRES AND CONDUITS (cont.)

WIRE CURRENT CAPACITY

Fahrenheit	140°F	167°F	194°F	140°F	167°F	194°F
Celsius	60°C	75°C	90°C	60°C	75°C	90°C
AWG	Group 1	Group 2	Group 3	Group 4	Group 5	Group 6
	COPPER			ALUMINUM		
18	N/A	N/A	14	N/A	N/A	N/A
16	N/A	N/A	18	N/A	N/A	N/A
14	20	20	25	N/A	N/A	N/A
12	25	25	30	20	20	25
10	30	35	40	25	30	35
8	40	50	55	30	40	45
6	55	65	75	40	50	60
4	70	85	95	55	65	75
3	85	100	110	65	75	85
2	95	115	130	75	90	100
1	110	130	150	85	100	115

Size						
1/0	125	150	170	100	120	135
2/0	145	175	195	115	135	150
3/0	165	200	225	130	155	175
4/0	195	230	260	150	180	205
250	215	255	290	170	205	230
300	240	285	320	190	230	255
350	260	310	350	210	250	280
400	280	335	380	225	270	305
500	320	380	430	260	310	350

Note: The above table is based on 3 wires in cable, ambient temp 86°F (30°C).

Wire Types:

Group 1: UF, TW

Group 2: RH, RHW, THW, THWN, XHHW, THHW, USE, FEPW, ZW

Group 3: THWN-2, XHH, USE-2, TA, TBS, SA, THHW, SIS, RHH, THW-2, THHN, XHHW, RHW-2, XHHW-2, ZW-2, FEP, MI

Group 4: UF, TW

Group 5: RH, RHW, THW, THWN, XHHW, THHW

Group 6: THWN-2, XHH, USE-2, TA, TBS, SA, THHW, SIS, RHH, THW-2, THHN, XHHW, RHW-2, XHHW-2, ZW-2

ELECTRICAL WIRES AND CONDUITS *(cont.)*		
VOLTAGE DROP TABLE		
Copper Conductor — 90% Power Factor		
AWG	**1"**	**3"**
14	0.4762	0.417
12	0.3125	0.263
10	0.1961	0.168
8	0.125	0.109
6	0.0833	0.071
4	0.0538	0.046
3	0.0431	0.038
2	0.0323	0.028
1	0.0323	0.028
0	0.0269	0.023
0	0.0222	0.02
0	0.019	0.016
0	0.0161	0.014
250	0.0147	0.013
300	0.0131	0.011
350	0.0121	0.011
400	0.0115	0.009
500	0.0101	0.009

ESTIMATING LABOR UNITS

SITE WORK

Items	Unit	Labor Hours
Site Demolition		
Clear and Grub	Acre	12 to 64
Tree removal	EA	2.5 to 7.0
Entire Building Demolition	SF	0.1 to 0.2
Paving Removal	SY	0.1 to 0.5
Debris Removal	CY	0.6 to 0.8
Building Interior Demo		
Demolishing footings	LF	0.1 to 0.3
Demolishing foundation wall	SF	0.1 to 0.2
Demolishing concrete slab	SF	0.05 to 0.10
Removing floor finish	SY	0.1 to 0.3
Demolishing wood framed wall	SF	0.05 to 0.10
Demolishing masonry block walls	SF	0.1 to 0.2
Demolishing ceiling	SF	0.02 to 0.05
Removing doors and frames	EA	0.5 to 1.5
Removing window and frame	EA	0.3 to 1.0
Demolishing roof structure and finish	SQ	2.0 to 5.0
Earthwork		
Excavation by hand	CY	1.1 to 2.3
Excavate trench by hand	LF	0.5 to 1.3
Moving excavated materials by hand	CY	1.0 to 2.0
Backfilling by hand	CY	0.4 to 0.7
Spreading soil piled on site by hand	CY	0.2 to 0.4
Paving		
Concrete Curbs	LF	0.2 to 0.3
Driveway Paving	SY	0.2 to 0.3
Asphalt Paving	SY	0.1 to 0.2
Hardscape Pavers	SY	1.3 to 2.1

EA: Each, CY: Cubic Yard, LF: Linear Foot, SF: Square Foot,
SQ: Square(100 square feet), SY: Square Yard

ESTIMATING LABOR UNITS (cont.)

CONCRETE

Items	Unit	Labor Hours
Formwork		
Footers	SFCA	0.1 to 0.2
Pads	SFCA	0.1 to 0.2
Foundation Walls	SFCA	0.2 to 0.5
Slabs	SFCA	0.8 to 1.2
Beams	SFCA	0.1 to 0.5
Columns	SFCA	0.1 to 0.2
Placing Concrete		
Footers	CY	0.3 to 0.8
Pads	CY	0.6 to 1.3
Foundation Walls	CY	0.4 to 0.9
Grade Beams	CY	0.3 to 0.7
Slab on Grade	CY	0.3 to 0.6
Elevated Slab	CY	0.4 to 0.8
Beams	CY	1.0 to 1.8
Columns	CY	0.3 to 1.3
Columns	CY	0.3 to 1.3
Placing Reinforcing Steel		
Footers	Ton	9 to 16
Pads	Ton	9 to 16
Foundation Walls	Ton	8 to 11
Slabs	Ton	13 to 15
Beams	Ton	12 to 21
Columns	Ton	13 to 22
Finishing Concrete		
Slab on Grade	SQ	1.0 to 3.5
Walls	SQ	3.0 to 3.5
Stairs	SQ	3.0 to 4.0
Curbs	SQ	2.5 to 3.0
CY: Cubic Yard, SFCA: Square Foot Contact Area, SQ: Square (100 square feet)		

ESTIMATING LABOR UNITS *(cont.)*			
MASONRY			
Items	Unit	Labor Hours	
		Mason	*Laborer*
Foundation CMU Wall			
4" thick	SQ	3.5 to 5.5	3.0 to 5.5
6" thick	SQ	4.0 to 6.0	3.5 to 6
8" thick	SQ	5 to 7	4.5 to 7.5
10" thick	SQ	6 to 9.5	7 to 11
12" thick	SQ	7 to 10	8 to 12
Exterior CMU Wall (up to 4' high)			
4" thick	SQ	3.5 to 5.5	3.5 to 6.0
6" thick	SQ	4.0 to 6.0	4.0 to 6.5
8" thick	SQ	4.5 to 6.0	5.0 to 7.5
10" thick	SQ	6.0 to 9.0	7.0 to 10.5
12" thick	SQ	7.0 to 10.0	8.0 to 11.5
Exterior CMU Wall (4' to 8' high)			
4" thick	SQ	3.5 to 6.0	4.5 to 7.5
6" thick	SQ	4.0 to 6.5	4.5 to 7.0
8" thick	SQ	4.5 to 6.5	6.0 to 9.0
10" thick	SQ	6.5 to 10.0	7.5 to 12.0
12" thick	SQ	7.5 to 10.0	8.5 to 12.0
Exterior CMU Wall (8' high and above)			
4" thick	SQ	4.5 to 8.0	6.0 to 9.5
6" thick	SQ	5.0 to 9.0	7.0 to 10.0
8" thick	SQ	5.0 to 7.0	7.0 to 10.0
10" thick	SQ	7.0 to 10.5	7.5 to 12.0
12" thick	SQ	7.5 to 10.0	8.5 to 12.0

ESTIMATING LABOR UNITS *(cont.)*			
MASONRY			
Items	Unit	**Labor Hours**	
		Mason	*Laborer*
Interior Block Wall			
4" thick	SQ	3.0 to 6.0	3.5 to 7.0
6" thick	SQ	3.5 to 6.5	4.5 to 7.5
8" thick	SQ	4.5 to 6.0	5.0 to 7.5
Face Brick (Standard/Modular)			
Common Bond	SQ	10.0 to 15.0	11.0 to 15.5
Running Bond	SQ	8.0 to 12.0	9.0 to 13.0
Stack Bond	SQ	12.0 to 18.0	10.0 to 15.0
Flemish Bond	SQ	11.0 to 16.0	12.0 to 16.0
English Bond	SQ	11.0 to 16.0	12.0 to 16.0
Face Brick (Oversize)			
Common Bond	SQ	12.5 to 19.0	14.0 to 19.5
Running Bond	SQ	12.0 to 15.0	11.5 to 16.5
Stack Bond	SQ	15.0 to 22.5	12.5 to 19.0
Flemish Bond	SQ	14.0 to 20.0	15.0 to 20.0
English Bond	SQ	14.0 to 20.0	15.0 to 20.0
SQ: Square (100 square feet)			

ESTIMATING LABOR UNITS *(cont.)*

STEEL

Items	Unit	Labor Hours
Structural Metal Framing		
Residential	Ton	7.8 to 9.3
Commercial/Institutional	Ton	7.8 to 9.3
Industrial	Ton	6.2 to 7.6
Steel Joists		
K Series	Ton	4.7 to 9.8
CS Series	Ton	5.3 to 8.9
LH, DLH and SLH Series	Ton	5.0 to 7.3
Joist Girders	Ton	5.3 to 7.3
Steel Decking		
22 to 14 Gage	SQ	1.8 to 3.5

CARPENTRY

Items	Unit	Labor Hours
Sills	1000 BF	19 to 40
Girders	1000 BF	8 to 33
Joists	1000 BF	16 to 24
Walls	1000 BF	18 to 30
Rafters	1000 BF	19 to 35
Decking	1000 BF	13 to 20
Subflooring	1000 BF	12 to 20
Timber framing	1000 BF	10 to 24
Wall Sheathing	1000 SF	12 to 20
Trim	100 LF	3 to 5
Framing for door/window opening	EA	1.5 to 3.5
Install Steel Doors	EA	1.5 to 2.5
Install Wood Doors	EA	1.0 to 2.0
Install Windows	EA	2.0 to 4.0

ESTIMATING LABOR UNITS *(cont.)*
ROOFING AND FINISHES

Items	Unit	Labor Hours
Roofing		
Asphalt Shingle	SQ	2.0 to 3.4
Wood Shingle	SQ	2.5 to 4.5
Clay Tile	SQ	4.5 to 6.0
Concrete Tile	SQ	5.5 to 6.0
Slate	SQ	4.8 to 11.5
Metal	SQ	3.0 to 4.5
Built-Up		
2 Ply	SQ	1.3 to 2.3
3 Ply	SQ	1.5 to 2.3
4 Ply	SQ	1.7 to 2.6
5 Ply	SQ	1.8 to 2.8
Aggregate Surface	SQ	0.3 to 0.5
Siding	SQ	2.0 to 4.0
Drywall		
Gypsum board on wall	SQ	2.0 to 4.0
Gypsum board on ceiling	SQ	3.0 to 4.5
Metal Framing	SQ	1.5 to 3.5
Flooring		
Wood	SQ	3.5 to 7.0
Vinyl	SQ	1.5 to 3.5
Carpet	SQ	2.5 to 3.5
Ceramic	SY	2.0 to 3.5
Terrazzo	SY	2.0 to 5.0
Acoustical ceiling	SQ	1.5 to 2.0
Stucco	SQ	5.5 to 8.0
Painting	SQ	0.5 to 1.5
Wall Paper	SQ	1.0 to 3.0

CHAPTER 7
Trade Estimating Forms

Standard forms help in the estimating process. Using forms or sheets is a good way to prepare a logical and consistent estimate. They do not guarantee that everything needed to be taken off will be covered. Contract documents and job conditions should be carefully reviewed first before using estimating forms.

In this chapter, estimating forms are furnished for your convenience. Two generic forms are introduced first to help in calculating quantities and subsequent pricing. Then specific quantity take-off forms are included for individual trades such as excavation, concrete, masonry, steel, carpentry, roofing, drywall, stucco, ceiling, flooring and paint, etc.

These blank quantity take-off forms can be photocopied for repeated use, or even customized to suit special needs such as for computerized estimating programs. If you have questions regarding how estimating forms work, what to take-off in each trade, or how to calculate the quantities, please refer to previous chapters for more information.

GENERAL TRADE QUANTITY TAKE-OFF FORM

Project:

Location

Trade:

Date:

Page ____ of ____

Take-off by:

| Item Description | Details | | | | Extension |
	Length	Width	Height	Count	
Total					

Note: Refer to previous chapters for more information.

7-2

GENERAL TRADE PRICING FORM

Item	Qty.	Unit	Material U/P	Material Subtotal	Labor U/P	Labor Subtotal	Item Subtotal
Subtotal							
Sales Tax							
Labor Burden							
Equipment							
Tools							
Supervision							
Other Overhead							
Profit							
Permit							
Total Price							

7-3

EXCAVATION QUANTITY TAKE-OFF FORM

Project: _____

Location: _____

Take-off by: _____

Date: _____

Page: ____ of ____

Checked by: _____

Description	Excavation Length (LF)	Excavation Width (LF)	Excavation Depth (LF)	Excavation Slope	Excavation Volume (CY)
Total					

Note: Refer to previous pages, including quantity-takeoff checklists, for more information.

7-4

CONCRETE CONTINUOUS FOOTING QUANTITY TAKE-OFF FORM

Project: _____

Location: _____

Take-off by: _____

Date: _____

Page: _____ of _____

Checked by: _____

Description	Footing Length (LF)	Footing Width (LF)	Footing Depth (LF)	Excavation Depth (LF)	Excavation Volume (CY)	Formwork (SFCA)	Concrete Volume (CY)	Rebar Weight (LB)	Backfill Volume (CY)
Total									

Note: Refer to previous pages, including quantity-takeoff checklists, for more information.

CONCRETE SPREAD FOOTING QUANTITY TAKE-OFF FORM

Project:							Date:			
Location:							Page: ____ of ____			
Take-off by:							Checked by:			

Description	No. of Pads (EA)	Pad Length (LF)	Pad Width (LF)	Pad Depth (LF)	Excavation Depth (LF)	Excavation Volume (CY)	Formwork (SFCA)	Concrete Volume (CY)	Rebar Weight (LB)	Backfill Volume (CY)
Total										

Note: Refer to previous pages, including quantity-takeoff checklists, for more information.

CONCRETE SLAB QUANTITY TAKE-OFF FORM

Project: _____

Location: _____

Take-off by: _____

Date: _____

Page: ____ of ____

Checked by: _____

Description	No. of Slabs (EA)	Slab Length (LF)	Slab Width (LF)	Slab Thickness (IN)	Slab Area (SF)	Formwork (SFCA)	Concrete Volume (CY)	Rebar Weight (LB)	Wire Mesh (Roll)
Total:									

Note: Refer to previous pages, including quantity-takeoff checklists, for more information

7-7

CONCRETE WALL QUANTITY TAKE-OFF FORM

Project: _____

Location: _____

Take-off by: _____

Date: _____

Page ___ of ___

Checked by: _____

Description	Wall Length (LF)	Wall Height (LF)	Wall Thickness (IN)	Wall Area (SF)	Formwork (SFCA)	Concrete Volume (CY)	Rebar Weight (LB)
Total							

Note: Refer to previous pages, including quantity-takeoff checklists, for more information.

CONCRETE BEAM QUANTITY TAKE-OFF FORM

Project: _____

Location: _____

Take-off by: _____

Date: _____

Page ____ of ____

Checked by: _____

Description	Beam Length (LF)	Beam Width (IN)	Beam Height (IN)	Formwork (SFCA)	Concrete Volume (CY)	Rebar Weight (LB)
Total						

Note: Refer to previous pages, including quantity-takeoff checklists, for more information.

CONCRETE COLUMN QUANTITY TAKE-OFF FORM

Project:			Date:			
Location			Page ___ of ___			
Take-off by:			Checked by:			

Description	Column Height (LF)	Section Width (IN)	Section Length (IN)	Formwork (SFCA)	Concrete Volume (CY)	Rebar Weight (LB)
Total						

Note: Refer to previous pages, including quantity-takeoff checklists, for more information.

MASONRY QUANTITY TAKE-OFF FORM

Project: _____

Location: _____

Take-off by: _____

Date: _____

Page ____ of ____

Checked by: _____

Description	Wall Length (LF)	Wall Height (LF)	Material Type	Block/Brick (EA)	Cell-fill Concrete (CY)	Rebar Weight (LB)
Total						

Note: Refer to previous pages, including quantity-takeoff checklists, for more information.

7-11

STRUCTURAL STEEL QUANTITY TAKE-OFF FORM

Project: _____

Location: _____

Take-off by: _____

Date: _____

Page _____ of _____

Checked by: _____

Reference	Shape Designation	No. of Shapes (EA)	Shape Length (LF)	Shape Unit Weight (LB/LF)	Shape Weight (LB)
Total					

Note: Refer to previous pages, including quantity-takeoff checklists, for more information.

STEEL DECKING QUANTITY TAKE-OFF FORM

Project:		Date:					
Location		Page ___ of ___					
Take-off by:		Checked by:					

Reference	Gauge	Deck Length (LF)	Deck Width (LF)	Overhang (LF)	Pitch	Lap (LF)	Deck Area (SF)
Total							

Note: Refer to previous pages, including quantity-takeoff checklists, for more information.

LUMBER QUANTITY TAKE-OFF FORM

Project:				Date:			
Location:				Page ___ of ___			
Take-off by:				Checked by:			

Description	Grade	QTY (EA)	Length (LF)	Section Dimension (IN x IN)	Extension (BF)
Total					

Note: Refer to previous pages, including quantity-takeoff checklists, for more information.

PLYWOOD QUANTITY TAKE-OFF FORM

Project:

Location:

Take-off by:

Date:

Page ____ of ____

Checked by:

Description	Grade	QTY (EA)	Thickness (IN)	Length (LF)	Width (LF)	Extension (SF)	Sheets (EA)
Total							

Note: Refer to previous pages, including quantity-takeoff checklists, for more information.

ROOFING QUANTITY TAKE-OFF FORM

Project:			Date:			
Location:			Page ___ of ___			
Take-off by:			Checked by:			
Description	Length (LF)	Width (LF)	Slope	Extension (SF)	In SQ
Total					

Note: Refer to previous pages, including quantity-takeoff checklists, for more information.

7-16

DRYWALL QUANTITY TAKE-OFF FORM

Project:

Location:

Take-off by:

Date:

Page _____ of _____

Checked by:

Description	Length (LF)	Height (LF)	Layers (EA)	Wall Area (SF)	Boards (EA)
Total					

Note: Refer to previous pages, including quantity-takeoff checklists, for more information.

STUCCO QUANTITY TAKE-OFF FORM

Project:				Date:	
Location:				Page ___ of ___	
Take-off by:				Checked by:	

Description	Thickness (IN)	Length (LF)	Height (LF)	Wall Area (SF)	Mortar (CF)
Total					

Note: Refer to previous pages, including quantity-takeoff checklists, for more information.

ACOUSTICAL CEILING QUANTITY TAKE-OFF FORM

Project:			Date:		
Location:			Page ___ of ___		
Take-off by:			Checked by:		

Description	Room Length (LF)	Room Width (LF)	Tile Size (IN x IN)	Ceiling Area (SF)	No. of Tiles (EA)
Total					

Note: Refer to previous pages, including quantity-takeoff checklists, for more information.

FLOORING QUANTITY TAKE-OFF FORM

Project:

Location:

Take-off by:

Date:

Page ____ of ____

Checked by:

Description	Flooring Type	Room Length (LF)	Room Width (LF)	Floor Area (SF)	Floor Base (LF)
Total					

Note: Refer to previous pages, including quantity-takeoff checklists, for more information.